IEE MANAGEMENT OF TECHNOLOGY SERIES 13

Series Editors: G. A. Montgomerie
B. C. Twiss

INTERCONNECTED MANUFACTURING SYSTEMS

the problems of advanced manufacturing

Other volumes in this series:

INTERCONNECTED MANUFACTURING SYSTEMS

the problems of
advanced manufacturing

H. Nicholson

Peter Peregrinus Ltd. on behalf of the Institution of Electrical Engineers

Published by: Peter Peregrinus Ltd., London, United Kingdom

Peter Peregrinus Ltd.,
Michael Faraday House,
Six Hills Way, Stevenage,
Herts. SG1 2AY, United Kingdom

British Library Cataloguing in Publication Data
Nicholson, Harold
   Structure of interconnected manufacturing systems.
   1. Manufacturing industries. Automation
   I. Title   II. Series
   670.427

ISBN 0 86341 224 6

Printed in England by Short Run Press Ltd., Exeter

Contents

Preface

This book provides an overview of the processes in discrete-event manufacturing systems, and emphasizes concepts relating to the interconnection of the wide range of activities involved. It highlights, particularly, the hierarchical structural properties of interconnection and the existence of forward (command) and reverse (monitoring) flows of information and activities which are encountered in most physical, biological and socio-economic systems.

The work thus focuses on concepts and generic models and does not attempt to provide a direct solution to the many complex problems encountered in Advanced Manufacturing Systems. It distills and reformulates material from the vast collection of other published work, and extends this to form a systems view of the complex large scale processes involved in manufacturing. In this way, it attempts to summarise briefly the salient features of automated manufacturing processes and to make them readily accessible to the reader requiring an overview of current developments.

Structurally, the book is divided into chapters essentially according to the hierarchical form of the advanced manufacturing system, moving from the higher-level strategic planning problems through product design, planning and scheduling to machine-level control and quality assurance. The aim is to present a fundamental body of knowledge relevant to the major activities in advanced manufacturing, including flexible manufacturing systems and computer integrated manufacturing, with particular emphasis on interconnected product and system design, planning, scheduling and control. It highlights principles related to the various activities and information flows involved, and although these may not be seen to be immediately relevant to the many small manufacturing systems in operation, they will need to be given serious consideration in future developments leading to the design, implementation and operation of automated manufacturing systems.

The text is suitable for undergraduate and postgraduate students engaged in the study of manufacturing systems and also for industrial engineers, designers and operators wishing to gain an overview of the many challenging problems which will be encountered.

Acknowledgments

The work has developed essentially from the author's desire to extend his previous interests in control systems into large scale system application areas. The stimulus for this was provided through contacts established with Dr. Peter Duff and Dr. Neil Sellors of the Mundella Office, University of Sheffield, during early 1987, and by discussions with teaching staff of the Stannington College of Further Education, now Loxley Tertiary College, Sheffield, including Dr. David Street and Mr. John Fallows, concerning the development of continuing education courses in manufacturing systems.

Useful discussions with staff of the Department of Mechanical and Process Engineering and the Department of Control Engineering, University of Sheffield, including Professors David Hayhurst and Bob Boucher, and Drs. Keith Ridgeway, Alan Morris, Paul Kidd and Gurvinder Virk, and particularly with Dr. Terence Perera of the Sheffield City Polytechnic, are also gratefully acknowledged. Other discussions with many industrialists and Mr. V. J. Osola, Executive Secretary of the Fellowship of Engineering, have also proved extremely useful and stimulating. Thanks are also due to Professor Bruno Maione of the University of Bari, Italy, for discussions and his kindness in sending me numerous references on Petri nets.

The author is particularly grateful to the University of Sheffield for the award of an Academic Development Fund grant which provided relief from some duties as Head of the Department of Control Engineering, and also for the subsequent award of a one year period of Study Leave during the Session 1987–88, without which this work would not have been completed successfully within a reasonable time scale.

The work has required the gathering and collation of information from numerous sources, which has provided the basis for many ideas and concepts, and hopefully these have been suitably acknowledged in the text. Apologies are extended for any omissions in this regard, and for possible misrepresentation of facts.

The author would finally like to express his appreciation to Mrs. Jose Stubbs for her assistance and very efficient typing of the manuscript.

Chapter 1

Introduction

1.1 General background

Manufacturing is concerned, in general terms, with the transformation of raw materials into finished products, and can be conceived as an input–output process with controllable variables (material, machines, transport devices, etc.), uncontrollable variables (market forces, customer demand, random disturbances etc.) and product outputs, as depicted in Fig. 1.1.

Within the advanced manufacturing system, activities will, essentially, link sequentially and involve:

- strategic planning at the company level,
- computer aided design for economic manufacture, and
- production design, planning, scheduling and control for component manufacture and assembly.

These will interrelate through material and component flows and communication channels, and effective operation of the overall complex will require the implementation of relatively advanced procedures for monitoring, planning and control, usually with multiple objectives and conflicts at all levels of decision making. The elements of a typical manufacturing process are illustrated in Fig. 1.2 (after [1]).

The complexity of the manufacturing system problem stems from the need to design, plan and control, often for large numbers of components (typically 20 000), operations per component (50), and work centres (300), with long lead times (18 months) and low machine utilisation (20 hours). This represents a large scale system problem, for which input-output relationships are generally lacking, and the analysis and design of the total system is difficult to perceive and formulate using a generic methodology.

There is difficulty in perceiving all the consequences of any given decision, and the problems cannot be solved without decomposing the system into a cellular structure involving small numbers of groups, parts, operations and machines, and possibly by reducing product variety. The problem is also complicated by uncertainty in customer demand, the current status of the system, supplier performance, and by machine variability and breakdowns.

Manufacturing is thus complex and multidimensional, and provides a rich source of challenging control and system-type problems [2]. Unfortunately, the range of existing control and OR theory does not provide directly applicable solutions, although the concepts of general systems theory, and particularly those involving the structural properties of large scale systems, can have important applications.

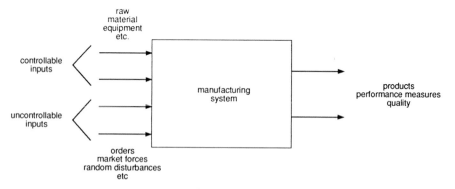

Fig. 1.1   Input–output representation of a manufacturing system

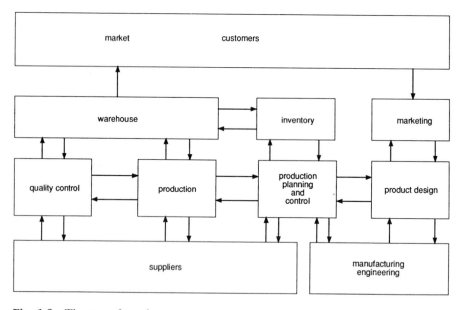

Fig. 1.2   The manufacturing process

1.1.1 Process classification
Production processes can be classified into two major categories—those producing materials and commodities in continuous processes, and manufacturing systems producing discrete parts or products. The present work emphasises the latter type of process, although many of the underlying concepts to be discussed are applicable to all production processes.

Manufacturing systems can be classified into: (*a*) job production, concerned with the manufacture of single items using general-purpose equipment, (*b*) high volume, flow-line production of a relatively small number of different products, using special-purpose equipment operating continuously over relatively long periods of time, and (c) batch production concerned with the manufacture of products in small or large batches or lots.

Batch production forms the major part of the Gross National Product of the UK and of other large producing countries, and the US Department of Commerce reports that 95% of all products in the United States are produced in lots of size 50 or less [3]. Many large scale operational systems, such as in road, rail and air transport, will also involve equipment requiring the production and assembly of many thousands of different parts [4].

1.1.2 Trends in batch manufacture

The trend in batch manufacture is towards high variety, high quality production, and the growing complexity and problems involved are now generally considered to require the support and flexibility of computerised automation of all activities from product design to the assembly and delivery of finished goods [5].

Automation is the key to survival for many companies requiring to operate competitively by producing high quality saleable goods to satisfy a fast changing market at the required time and at an acceptable cost. There is also a growing awareness of the need to integrate all company activities, from corporate strategy to product design, manufacture, quality control and delivery, in order to make optimum use of all resources and to produce a realistic return on investment. This requires effective planning and scheduling strategies, and if the overall system is properly organised it will enable batch items of wide variety to be produced at costs approaching those of volume production, with reduced inventories and manufacturing lead times [6].

System integration, however, is not easy and requires an acute awareness of input–output properties and of the effects of interaction which cannot in general terms be defined analytically, and of the effects of system changes, such as machine breakdowns, propagating throughout the overall system. It is also generally agreed that before proceeding with full-scale automation of an existing manufacturing system, the first consideration should be a detailed investigation of existing company practice—and the best initial strategy is to modify existing working arrangements in order to reduce work-in-process.

1.1.3 Problems of manufacturing

The design and operation of advanced manufacturing processes involves complex interactions between technological and management systems, and introduce very significant problems. These are increasingly becoming problems of planning involving the efficient co-ordination of design and manufacturing, material supply and handling, inventory management, quality control and zero fault production. Technical improvements in equipment technology have advanced significantly, but there remains the need to develop generalised approaches to the integration of decision making processes and information flows.

Experience has shown that problems and failures can arise through

- lack of employee involvement in design and planning and of management commitment,
- inadequate planning of systems and management structures,
- insufficient staff training in the new technologies and in software requirements, and
- lack of company involvement with equipment and material suppliers.

Different management structures and the formation of task groups for team working are necessary for successful implementation of new technology. Communications must also be established between all planning and operational groups, and particularly between the closely related activities of computer aided design (CAD) and computer aided manufacture (CAM).

Manufacturing industry will be installing advanced automated systems at an increasing rate during the next few years [7]. Investment by the smaller companies has, however, been restricted, probably due to a lack of understanding of the technical and economic issues involved, and also because there is no simple solution for achieving profitability. There are also indications from Japanese practice that the use of advanced technology may not be an essential prerequisite for profitable manufacture. The use of the Japanese Kanban and Just-in-Time (JIT) methods have been very successful without the use of advanced integrated automation systems and complex planning techniques. These have produced improvements in productivity and machine utilisation by reducing work-in-process and high inventories of unfinished parts on the shop floor. They have also established a significant trend towards the implementation of 'zero inventory'-type methods and their integration with other computer-based planning and control methods.

Developments in Integrated Manufacturing Systems are, however, proceeding rapidly, and the general consensus is that these are essential for survival in highly competitive world markets, and for progress towards the fully 'automated factory' of the future.

1.2 Recent historical developments

Mechanisation and automation in manufacturing industry has increased significantly since the introduction of transfer lines during the early 1940s and Numerically Controlled (NC) machines during the 1950s. A major development was the demonstration of an NC machine by the Massachusetts Institute of Technology in 1952 which initiated a new era in manufacturing [3]. Computer Numerically Controlled (CNC) machines then developed during the 1960s, and have since been designed with automated handling devices and robots and tool-wear compensation systems [8]. Direct Numerical Control (DNC) technology providing central supervision of a number of loosely connected machines then followed, which with the development of programmable logic controllers and microcomputers made possible the grouping of CNC machines into a semi-autonomous cell arrangement for manufacturing a range of components [7].

The cell configuration formed the nucleus for the development of larger-scale flexible manufacturing systems (FMSs) or programmable job shops, consisting of computer controlled machining centres or cells linked by an automated materials handling system. FMSs are designed to machine different part types and produce groups of components, and to respond dynamically to changing operating conditions such as unpredictable demand patterns and machine failures. They are expected to utilise machine tools and material handling equipment more effectively, thereby reducing material stocks, work-in-process and manufacturing delays, and increasing process output. Their application is a necessary step towards improved productivity in many batch manufacturing industries.

Considerable potential exists for the introduction of FMSs in companies with medium-sized production runs (200–20 000 parts per year), which account for 60 to 80% of the value of all parts manufactured [9], [10]. Investment has, however, been restricted to a relatively small number of companies worldwide, again due to the difficulties in quantifying economic benefits and because of the lack of trained staff with technical knowledge and understanding of FMS capabilities which need to be matched with strategic planning and marketing. Successful applications are to be found in large manufacturing companies, such as in the aircraft industry, producing large numbers of complex components with a relatively high rate of design change.

The next stage of development incorporating and extending the concepts of flexible manufacturing was and continues to be Computer Integrated Manufacturing (CIM), using a hierarchy of computer controlled subsystems and knowledge based decision processes to effect complete integration. The CIM initiative is intended to extend existing schemes of automation, concerned, for example, with product CAD, capacity planning, material requirements planning, scheduling and inventory control, by integrating these activities with computer aided strategic planning, automated material handling, real-time machine control, automated assembly and quality assessment and control, through to marketing and sales.

Whilst the trend to planned integration in CIM is well advanced, particularly at the research and development level, there are major problems in implementing the overall concepts, mainly because of the lack of hardware/software standards required to interconnect multi-vendor equipment. Problems also arise through the difficulties of establishing combined product design–manufacturing data bases incorporating facilities for activating discrete-event state models associated with multi-level scheduling and real-time adaptive control. There is also a lack of understanding of the organisational problems involved in implementing a different philosophy of management control required with extended integration.

1.2.1 Production control

It is interesting to note that the concept of integration in production control originated much earlier [11—13] than the recent CIM developments for batch manufacture. Also, main-frame computer programs for bills of materials, process planning and shop floor control had been implemented previously, during the 1970s [14e].

Integration in large scale process control was being undertaken in the UK, at the Park Gate Iron and Steel Company, Rotherham, during the mid 1960s.

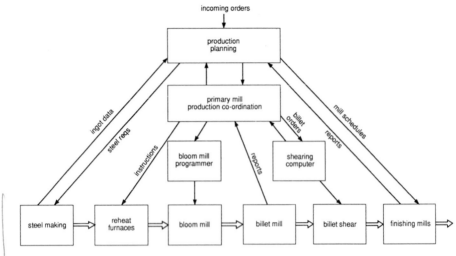

Fig. 1.3   Integrated computer control system

This included production planning and overall control of a steel process from the receipt of orders to the dispatch of finished products [12], using a hierarchy of computers as illustrated in Fig. 1.3 (after [15]). The upper level computer performed off-line administration and production scheduling, including the issuing and receipt of orders and reports from various process departments. The lower level computers co-ordinated on-line production control, by identifying and tracking material flow and preparing working instructions to achieve steady-state optimised control of the rolled billet lengths. The system produced a wide range of steel qualities and parcel sizes, and was effective in smoothing production and increasing efficiency.

Research was also being undertaken, as early as 1968 [16b], into complete automated manufacturing, utilising CAD, automated planning, inventory control, automated inspection devices, and partially automated decision making. Proposals were made to link CAD to manufacture based on the selection of a part family of components with common characteristics. The matching of machine tool–workpiece requirements were to be achieved using a machine characteristics matrix of machine data such as types of operations, maximum horse power, tool capacity etc., and a tool characteristics matrix including tool types, tool life parameters etc. The process was to be planned and optimised at the completion of the tool selection phase using a mathematical model for a particular part family, and a criterion of minimum time. Fig. 1.4 illustrates the conceptual basis of this ambitious programme.

A research initiative in Integrated Production Systems was later proposed by the UK Control Engineering Committee of SRC (now SERC—Science and Engineering Research Council) in 1979. Broad areas of research were designated, including the design of data bases and communication systems, decision making, control system structures and integration, and human factors. Discussions with industrialists revealed the need to design small-scale flexible

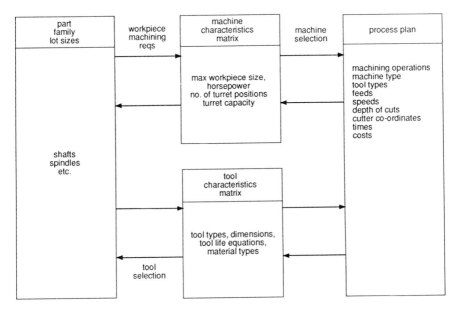

Fig. 1.4 Machine tool–workpiece requirements matching

plant for integrated system control, reduced response times, and to investigate information flows within management structures. A Manufacturing Technology Committee, in discussing research into the Efficiency of Batch Manufacture (1978), also highlighted possible research areas including the economic loading of high capacity machines, the need for more disciplined demand scheduling, the shortening of delivery times, and improved stocking policies, forward scheduling and production planning.

It is of interest to note that elements of the present-day CIM initiatives are indicated clearly in these previous developments, although in some cases the missing factor was computer aided support for product design, planning and manufacture.

1.3 System components

1.3.1 Machine tools

An automated manufacturing system will include NC machines with dedicated logic control or CNC machines with software control, for turning, boring, milling, grinding, cutting, punching etc. They will often have versatile pro-grammed facilities for sequencing and positioning of tool operation with tool and speed selection and servo-motor control of tool position and cutting rate. The manufacturing system may also contain more universal machining centres incorporating multiple-spindle machines with automatic tool head changers and tool buffers, with program-controlled tool changing and selection [10]. Automatic tool changing facilitates the combination of many conventional operations, and reduces the amount of work handling and consequently the work-in-process and lead times.

DNC computer systems are used to schedule and supervise several NC machine tools directly, by co-ordinating and synchronising the storage and distribution of workpiece data and the execution of programs required with changing products. This facilitates the integration of machine tool control with process planning, material handling and assembly, using an external process computer. Advanced systems will also incorporate automatic tool changing and computerised tool selection, to support the on-line scheduling of machine capacity, which is of particular benefit in reducing the effects of equipment failure by machine re-scheduling. Machine tools may also be served by robots, and incorporate automated inspection equipment.

1.3.1.1 Tool systems.   Machine tools can represent a large proportion of the total investment in a manufacturing system, of the order of 10% [17a], and thus must be used efficiently. Each tool may be characterised by many characteristics, such as tool type, class, offsets, cutting life etc., and a medium size flexible manufacturing system may require a relatively large number of tools, possibly of the order of thousands [17b], which introduces complex problems of tool handling and scheduling.

A tool management system is required to provide the correct tools at the right place at the right time, and to reduce unnecessary tool changes. Monitoring of tool location and tool life will be required to ensure that tools for a particular operation are present, and that accumulated wear for each tool will allow sufficient tool life for completion of the operation [18a]. The on-line measurement of tool wear is usually desirable, and inferential measurement techniques may use, for example, a spectral analysis of spindle power requirements [19] or vision systems for detecting shape and tool absence. These could then provide the means for implementing closed loop adaptive machine tool control to extend tool life, maintain surface quality and maximise production. Facilities might also be provided for the automatic correction of machine distortion, spindle and measurement errors, and for the optimisation of tool trajectories.

1.3.2 Material handling systems

The effective movement of materials and reduction of in-process inventory and stocks, consistent with efficient production and customer demand, are essential for achieving the full benefits of integrated manufacturing. Materials handling and storage accounts for a relatively high proportion of the residence time of material and components from entry to completion and delivery of the finished product, and can represent a high proportion (30–40%) of manufacturing costs [20]. The material handling system can also very often cause bottlenecks, and its design should be closely related to the work-in-process management policy and storage requirements [21].

1.3.2.1 Trends in material handling devices.   *Material transport mechanisms* can be divided into the following categories [22, 23]:

(a) *Powered carriers*—including, automatically guided vehicles (AGVs) tracking a floor guide wire, power-and-free conveyors, tow-carts, and rail-guided vehicles

(b) *Conveyors*—for transporting carriers, include powered roller, belt, air-cushion and modular conveyors

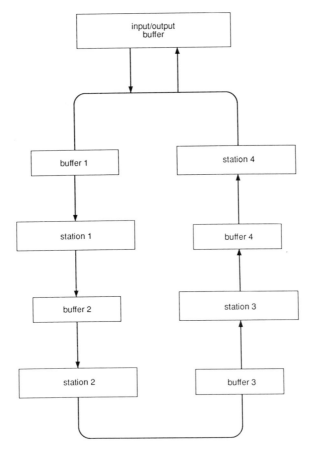

Fig. 1.5 Serial-loop routing pattern

Floor traffic patterns for a material handling system serving a basic layout of stations and buffers can take the following forms:

(a) *Single-loop pattern*—Fig. 1.5 illustrates the form of a simple serial-loop pattern serving a single sequence of processes, with carriers loaded into and removed from the system at the input/output buffer (after [22]). Routing between stations is through adjacent buffers and stations. Blockages can occur with preceeding carriers awaiting processing, and the arrangement is not desirable in a FMS. Processing at more than one station also requires multiple loop traversing.

(b) *Single-recirculating-loop pattern*—Many of the above problems can be eliminated using a circulating loop system with spur connections, as illustrated in Fig. 1.6. Each station and buffer have a separate subloop which allows carriers to bypass the station, and the main loop will have capacity to hold buffering stock in circulation until required. Traverse times can, however, become excessive with many stations and long loops, particularly if parts

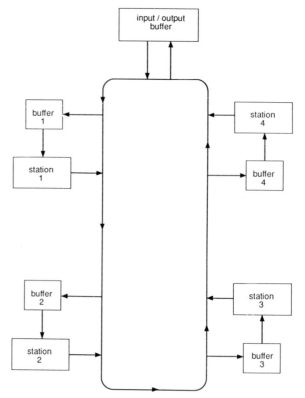

Fig. 1.6   Single-recirculating-loop routing pattern

have to visit more than one station out of sequence. This increases lead times, and excessive work-in-process inventory may be required.

(c) *Multiple-recirculating-loop pattern*—The above disadvantages can be reduced with the traffic configuration illustrated in Fig. 1.7, which includes smaller, multiple recirculating loops and more direct paths between stations, and provides increased routing flexibility. The spur connections give independent access to both stations and buffers, and multiple paths exist between all stations. A complex routing algorithm will, however, be required to locate all possible routes through the system, and selected routing will require the use of a pre-stored system map.

A general real-time, logic-routing algorithm appropriate to the loop system of Fig. 1.7 has been developed [22], allowing the process stations to be visited in any required order, and routing decisions to be made at intersections, without global knowledge of the system layout.

The loading and unloading of palletised components on machining centres will require high positioning accuracy [3] and the correction of pallet alignment and work mounting errors, and the problem has been investigated using theoretical analysis [18b]. This involves the measurement of the loaded pallet

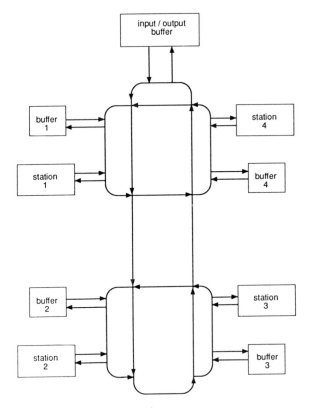

Fig. 1.7 Multiple-recirculating-loop routing pattern

position and calculation of the average pallet positioning and orientation error vector for any pallet and machine combination, and the automatic compensation for some of the known errors by the cell controller.

1.3.3 Robots

Industrial robots have important applications in automated manufacturing as mechanical manipulators. They have multiple roles, particularly in the work-station area, for palletising, machine loading, tool changing, component transfer, interstage loading, deburring, arc welding, and automated parts assembly. Their applications in warehouse automation include depalletisation, item picking and collation.

Standards for robot design and operation, including guarding and interlocking, positioning performance, programming and interface requirements, have been developed or are under investigation [18c].

The problem of robot failure recovery has been considered [24a] by 'viewing the execution of a task as a series of transformations from an initial state to a goal state via a series of predetermined intermediate states'. Failure then represents the occurrence of an unexpected state. The error recovery problem is

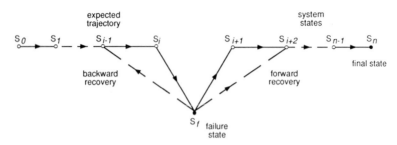

Fig. 1.8   Task as a sequence of state transformations with fault occurrence

concerned with reaching a goal state from the current, failure state, as depicted in Fig. 1.8 (after [24*a*]). Error detection could be related to discrepancies occurring between a sensor based 'world' model of the state of the robot and sensory expectations.

1.3.3.1 Automated assembly.   Assembly accounts for a large proportion of production time (53% is cited) [25*d*], and is one of the most difficult manufacturing operations to automate. In robot-based product assembly, involving the feeding, fitting and joining together of parts, very high precision movements are required, together with accurate sensing and information processing, in contrast to their applications in continuous processes such as welding [27].

Further developments are required, particularly for part location and orientation, and parts must be designed suitable for automated assembly. There must also be developments of advanced sensing such as vision systems, universal grippers with touch sensitivity and versatile feeders with parts presentation. Future robots will also need to incorporate capabilities in pattern recognition, artificial intelligence and voice response [28].

Current robotics research has been attempting to improve the flexibility of assembly cells and the ability to adapt to less accurate motion specifications, in order to reduce the number of defective rejects and to detect impending collisions. Robots must be developed which will operate not according to set programs of instructions but which have the ability and 'intelligence' to generate adaptive movement control, in order to respond to unexpected events such as handling different shapes of components. They must be able to interact with other robots, and be aware of their movements, and also with the external unstructured environment, and be able to interpret time, using on-line dynamic vision sensor tracking and touch mechanisms [7]. The vision system would guide the robot in handling touching parts, with the tactile system providing spatial correction data.

The analysis of assembly operations shows that trajectory control with quasi-continuous feedback and state surveillance using discrete 'status' information, can reduce the effects of inaccurate motion specifications [27]. The assembly operation is also not simply reducible to a computational sequence—it is more a logical problem-solving activity requiring intelligence.

The natural description of product assembly tasks has been considered in terms of motions terminated by sudden change of force conditions, and vice versa [30]. Assembly described only in terms of positionally controlled motion is considered unnatural for most product assembly, and is only possible if component part tolerances can be maintained at a level necessary for assembly. Commercially available industrial robots are usually low-tolerance devices, and require to be force controlled to be suitable for assembly of a large class of products.

Other developments consider the design of a self-tuning-type adaptive controller for position and velocity control of manipulator joints [31]. The motion of the joints is modelled using an autoregressive time series model, and model parameters are determined using an on-line recursive algorithm. The controller will adapt to changing operating conditions; for example with the effective inertia of a moving object changing along the trajectory of the joint positions.

The automation of complex handling and assembly operations using industrial robots will also require the robot to respond to emergency situations, such as when a defective component part is encountered. The lack of automatic emergency recovery is a serious limitation of present industrial robots, and solutions have been proposed using artificial intelligence techniques to provide recovery trajectories and desired re-entry from error conditions [32].

Chapter 2
System Structure

2.1 Introduction

Interconnected manufacturing systems possess a hierarchic structure similar to that of many other large scale technological and social systems. Hierarchic systems have generic properties that are independent of their specific context, and will usually be composed of interrelated subsystems, of few different kinds, each of which will be hierarchic in structure and have their own subsystems [33].

Higher-level, executive-type systems will usually be characterised by lower-frequency dynamics, longer planning horizons and aggregated data, and the lower-level subsystems by higher-frequency dynamics with shorter horizons and will generally use more detailed information – similar to activities in the Integrated Manufacturing System [34]. Intra-component connections will usually be stronger than inter-component linkages, which effectively separates the internal structure of the lower level subsystems from the higher levels. Hierarchies thus often possess the property of near-decomposability, and only aggregative properties of their parts enter into the description of the interactions of those parts. This greatly simplifies the representation and analysis of large scale systems.

It will usually be difficult and often impossible to model mathematically and perceive an integration of the total system, and the approach to understanding and synthesis has to be structured in terms of subsystem behaviour and knowledge of the interconnecting variables.

2.2 Manufacturing system structure

The structured activities and information flows in a manufacturing system will generally commence, in broad terms, with design and proceed to planning and control, as a sequential connection of subsystem operations. The associated flows of information and movements of material and parts through the process will require the generation of commands and their transmission in a 'forward' direction to the lower-level subsystems. The effects of flow changes in the process, produced for example by disturbances such as machine breakdown, will 'reflect' back through the coupled subsystems, and the monitoring of the effects of decisions made by higher-level subsystems required to achieve some measure of integrated control, will also require a 'backwards' transmission or

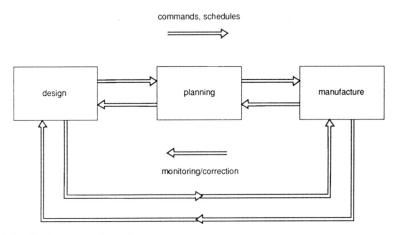

Fig. 2.1 Design–manufacturing system

feedback of information. These effects of interaction, inherent in the integrated manufacturing system, are illustrated in Fig. 2.1.

The concepts of a forward–reverse, feedback or scattering type of structure [35] will exist generally in all large scale, interconnected, physical, socio-economic and biological control systems, and can be used in principle to effect a functional decomposition of complex systems.

2.2.1 Hierarchical structure

The activities and information flows in the integrated manufacturing system form a hierarchical structure linking the strategic corporate management level with the factory level operations involving FMSs, cells and workstations. The hierarchy is characterised by a range of time scales or horizons associated with long-term decisions for capital expenditure and future planning at the corporate level based on demand forecasts and economic constraints (one to five years), medium-term decisions for production planning, order processing, stock control, machine allocation etc. (daily to monthly), and short-term decisions for machine tool control, component and material handling, robot control etc., at the machine level (seconds to hours). These different horizons and time scales effectively decouple the hierarchy into a sequence of autonomous subsystems.

The sequential coupling of the design, planning and production activities requires a hierarchical structuring of information, contained within or linked to a distributed data base. Each subsystem will receive commands as schedules and goals from the preceding higher level, and report back status information for performance monitoring and possibly for job rescheduling. The highest level generates long-range goals which are transmitted as shorter-range sequences of subtasks or plans leading to schedules or control at the lower levels. Each control system operates with local information and feedback, within the constraints and goals set by the higher level production plan [36]. Local decision making permits continued operation of the lower levels with higher-level failures and disturbances, and can accommodate rescheduling of material flow and machining operations following machine failures and bottlenecks. The higher levels will only need to intervene with correcting actions to the lower levels when their operations deviate from preprogrammed control levels. A typical hierarchical arrangement of subsystems incorporating planning, scheduling, order release, control and verification functions is outlined in Fig. 2.2 (after [10]), in a scattering-type format.

Requirements planning will be based on knowledge of orders, manufacturing resources and also a product description, possibly including a detailed design based on a functional representation, material properties, knowledge of manufacturing methods, costs etc. The output of the planning function subsystem initiates input scheduling incorporating the constraints of factory resources, and leads on to order release and bills of material to activate manufacturing control

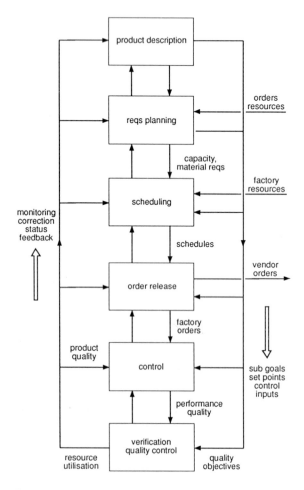

Fig. 2.2   Manufacturing system activities

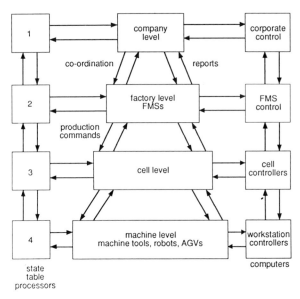

Fig. 2.3 Large scale manufacturing system hierarchy

at the workstation or machine level. Verification procedures including performance monitoring and quality control may also be integrated into the hierarchy to inject corrective feedback action into higher level activities.

The various broad levels of an integrated manufacturing system are indicated in Fig. 2.3, and typical management and plant control decision functions are detailed in Table 2.1. The control structure is represented as a hierarchy of feedback controllers implemented as state machines, and subsystem inputs, outputs and states can be represented in state graphs [37]. The table for each control subsystem is processed at regular intervals to identify the current state, which is used to generate appropriate outputs for maintaining the system within acceptable ranges.

2.3 Large scale system control

The theory of large scale system control has been studied extensively in the Control Systems field, and the underlying concepts relating, particularly, to multilevel-multigoal decentralised control and co-ordination, have important applications in manufacturing system control. A brief account of relevant work is included, and although this is based on continuous- and discrete-time model representations, the concepts will provide a framework for the development of a model-based methodology required for the study of discrete-event, multicommodity network flow problems in computer integrated manufacturing.

A formalism for the design of large scale system control based on the interconnection of multilevel and hierarchically co-ordinated subsystems was

Table 2.1 Typical management and plant control decision functions

---

Corporate/factory level
Strategic long-range planning, plant investment policy
Market research, sales forecasting
New products, specifications, quality standards
Plant layout, simulation
Financial planning, budgets, production costs, pricing strategies

---

Plant level
Product design—manufacturing requirements
Production planning and control
Order processing, bill of materials, material requirements
Resource scheduling, machine loading, line balancing
NC programs
Material movement, inventory control
Assembly scheduling
Quality control, test programs
Maintenance scheduling

Cell/machine level
Machine assignments, fixturing
Material and tool requirements, handling
Real-time machine tool and robot control, monitoring
Assembly
Inspection and quality control

---

developed during the early 1960s by Mesarovic and others [38—40] at the Systems Research Centre, Case Institute of Technology, USA. Decomposition was used to effect a division of control into separate lower-level 'infimal' controllers, and the interacting variables were taken into account by an upper-level 'supremal' controller, using an 'interaction prediction' or 'interaction balance' principle to co-ordinate the infimal units.

An extensive range of other multivariable system theory concerned with the control of multi-input, multi-output linear systems has also been developed, but its relevance and possible application to the overall control of large scale manufacturing systems is doubtful, except at the detailed machine level. The theory is restricted to relatively low order multivariable systems with a small number of inputs and outputs, and its application generally yields complex and 'dense' controller structures. It cannot be used to solve the combinational problems prevalent in manufacturing systems, which require the design of sparsely structured controllers reliable over wide ranges of operation and in failure modes.

Adaptive control strategies are also required with plant variability, although again design methods are not highly developed for large scale systems. New concepts and procedures are needed for designing simplified decision rules and reliable controllers for large scale systems with multi-resource scheduling and conflicting objectives, using efficient signal processing techniques with limited information and distributed data bases to exploit available information patterns and the structural properties of systems.

There is currently no cohesive body of knowledge, or method of application of control or operations research strategies, which will achieve the objectives of integrated control for large scale and complex manufacturing systems. New ideas and concepts are required for generating simplified sub-optimal decision rules which will integrate the range of processes involved, for efficient and cost-effective production.

2.3.1 Decentralised control

Most large scale systems, including integrated manufacturing, include a complex interconnection of equipment and information flows, and it may be difficult if not impossible to perceive the fundamental mechanisms involved and to design an integrated controller for such systems. The ability to design effective planning and control strategies will also depend critically on the availability of adequate system models, knowledge bases and realistic objective functions.

Large systems will need to be broken down for analysis into a number of interacting subsystems. Some systems will possess natural divisions, as with sequentially connected subsystems, whereas in other cases an arbitrary mathematical decomposition may be necessary, which should aim to produce minimum interactions between different sections [41]. A method of decomposition based on the tearing of links connecting a basic tree structure, as developed for the structured analysis of electrical networks [35], may also find possible application in the design of integrated manufacturing systems involving the control of discrete-event network flows.

Transport lags, produced for example by material movement, will effectively produce the effects of decoupling in sequential processes, and permit the application of local independent control with feed-forward action used to counteract the effects of the time delays. These concepts have direct relevance in manufacturing systems in which material movement and machining operations produce a cascade of time delays and also more general connections of delays and queueing effects, in discrete-event processes.

Decomposition will be more difficult in the general system case, in which each component will be mutually coupled with forward and reverse connections to a number of adjacent processes, as shown generally in Fig. 2.4 for a system containing five tightly coupled subsystems. It will be possible, however, to identify a pyramid-like structure of local decision problems and goals, within the hierarchical chain of command in most organisations, and particularly within the manufacturing system problem beyond the detailed machine level.

The philosophy of decentralised control for multi-level, multi-goal systems [38—41], in which the overall system is subdivided into a number of smaller subsystems operating with local controllers, is illustrated in the multi-level

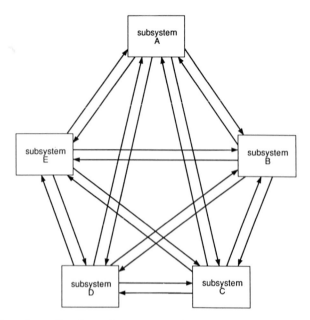

Fig. 2.4   Mutual coupling in an interconnected system

structure of Fig. 2.5. Each controller has access to only a limited set of system measurements and generates control action using a model-based local objective function. A higher-level co-ordinator with an overall objective function, and receiving information from the local controllers, can modify local control action and compensate for the overall effects of decentralised control and the errors and delays in the communication channels, and also for model inaccuracies and the possible loss of any local controller.

The overall system goal is achieved using higher levels of supervisory control with assigned goals to co-ordinate the decentralised lower-level control problems using intervention parameters set by the co-ordinators. The advantage of the structure is that any one controller is responsible for only a few subordinate controllers, and the overall system reliability will not be affected unduly by the actions of individual components. The multi-level, multi-goal concept is of central significance in the study of complex, large-scale systems. However, the development of realistic models, and the design of a co-ordinated approach to decentralised control of static and dynamic states in the presence of stochastic disturbances, will be a difficult task [42].

2.3.2 Decentralised manufacturing system control
The concepts of decentralised control have particular relevance to the sequential flow problems in manufacturing, in which a higher-level controller, say at the corporate, plant or cell level, may be designed to receive information from lower levels and issue goals and intervention parameters for subsystem control. A typical example exists with the co-ordinated control of the cell-machine tool-material handling system complex in a flexible manufacturing system, depicted

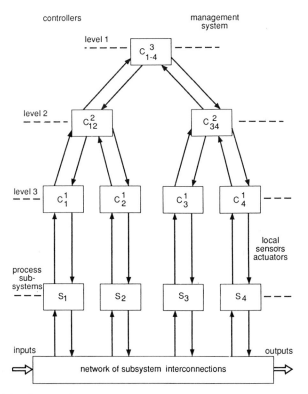

Fig. 2.5 Multi-level control structure

schematically in Fig. 2.6. The cell subsystem is incorporated within the hierarchy of the overall system, and itself includes a hierarchical structure which is decomposable into different temporal levels associated with the longer-term horizon of the material handling system and the higher frequency dynamics of the machine tools.

The local controllers may be involved in tracking a desired response or schedule, according to some objective criterion, using say a state machine to specify planned activities such as material movement. The controllers may also incorporate optimal (Kalman-type) filtering required with noisy measurements and for model parameter updating. With transport lags in the serial connections, and also with more general feedback loops and external disturbances, it is feasible in principle for the model-based supervisory controller to monitor and predict expected and required inputs and outputs in the forward processes, and to readjust local control action to achieve desired overall performance. In this way, the supervisory controller effectively reduces the effects of coupling between adjacent processes.

If a centralised information pattern or data base is available, in contrast to the decentralised local patterns assumed previously, modern control theory would also propose a two-level decentralised structure, using a combined open

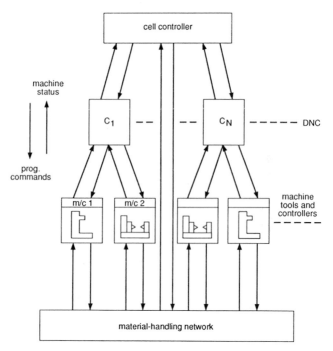

Fig. 2.6   Cell–machine tool–MHS complex

loop control at a 'strategic' level, derived using, say, nonlinear programming to optimise an objective function with a static model, and a tactical optimal stabilising control based for example on a Linear–Quadratic–Gaussian formulation using a dynamic model and Kalman filters. This would provide tracking of the desired response defined by the strategic level and corrections required due to the presence of stochastic disturbances and modelling errors at the strategic level. A structural configuration of this arrangement is illustrated in Fig. 2.7 (after [42]). The concepts are applicable within the manufacturing system, although the development of algorithmic solutions will require a reformulation of existing continuous-time model results into a discrete-event model framework, and further detailed work is required.

The extension of these concepts to larger scale systems leads to a hierarchical structure incorporating 'district' and 'area' controllers arranged similarly in a multi-level pyramid-type structure. Each higher level, with increasing time horizons, will predict responses and provide supervisory control over combined sections of lower level controllers. These ideas have been extended to the multi-level problems encountered in large scale production systems, and to the need to integrate the activities of marketing, accounting, production planning, material supplies, inventory and labour, into an overall, model-based control hierarchy. A proposed structure is illustrated in Fig. 2.8 (after [41]), and it is interesting to note a similarity to the concepts of Computer Integrated Manufacturing presently being investigated in the manufacturing systems field.

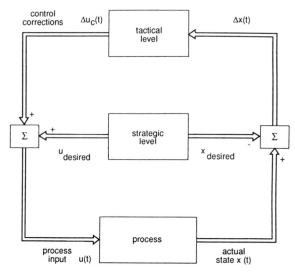

Fig. 2.7 Two-level decentralised control structure

2.4 Structured analysis

Structured analysis can be used as a means of representing and linking the vast array of activities and data flows existing in a large scale manufacturing complex. Company operations can be divided into independent subsystem models which can be used to simplify and rationalise company organisation and controls.

2.4.1 SADT model

The concept of functional modelling incorporating a structured decomposition technique is illustrated in Reference [8] with an application in batch manufacturing. This gives a useful account of the Structured Analysis and Design Technique (SADT)* of Ross [43] developed by SofTech Inc., and highlights the main features which involve a successive decomposition of activities and data flows to simple more detailed tasks. The method has been applied to a relatively wide range of design, planning and control problems, particularly in computer hardware/software systems.

The technique identifies activities and data flows within inter-connected block structures (usually three to six in number), with each block indicating a qualitative relationship between flows, constraints and mechanisms. The basic structure of the model blocks is illustrated in Fig. 2.9. The modelling technique is based essentially on the expansion of selected components of the aggregate form of the overall system model (A0(D0)), containing inputs, outputs, constraints and mechanisms, into more detailed interconnected representations at the next level of decomposition. The process of decomposition can be

* SADT is a registered trademark of SofTech Inc, Boston, Mass, USA

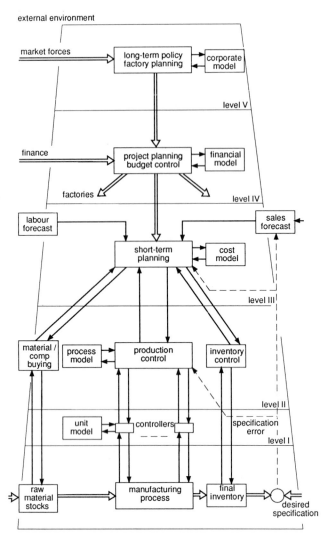

Fig. 2.8 Hierarchical production strategy

continued on to additional detailed levels, provided the inputs, outputs, constraints and mechanisms to any box are preserved in the expanded form at the next level. Fig. 2.10 (after [8]) illustrates the basic concept of decomposition. The method also involves forming a structured list identifying each activity and data box.

The graphical technique has been further extended in the US Air Force's IDEF suite of programs in the ICAM programme, which can be used to model the fundamental relationships (IDEF(0)), data requirements (IDEF(1)) and

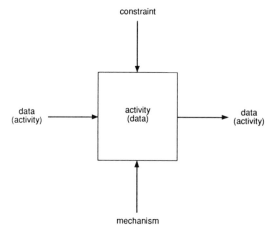

Fig. 2.9 Basic structure of an activity (data) model block

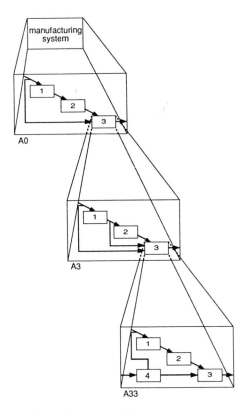

Fig. 2.10 Decomposition of system activities

the dynamic behaviour (IDEF(2)) of complex manufacturing systems [26a].*
Computer based structured design methods can have important applications in
manufacturing system design, although further developments are required to
incorporate facilities for measuring the quality of a given design and optimising
the specifications. The availability of enhanced structured design methods
incorporating means of configuring generic modules representing, for example,
strategic planning, operational planning, scheduling and control, material
handling and storage, within a network structure, will also contribute significantly to the development of computer integrated manufacturing systems.

2.4.1.1 Example of SADT   The following simple example, including a single
level of decomposition, illustrates the basic ideas of the structured analysis
technique. More detailed information relating to the structured decomposition
of a company engaged in the batch production of sheet-metal components is
given in Reference [8].

For the particular company, the activity node list includes statements such
as:

A0   RUN COMPANY
     A1   Formulate Strategy
        A11 . . . . .
          .
          .
          .

     A2   Formulate Company Plan
        A21 . . . . .
          .
          .
          .

     A3   Implement Company Plan
        A31   Manage Finance
        A32   Design Products and Tools
        A33   Manufacture Products
            A331   Prepare Production Programme
                A3311   Plan Production Methods
                  .
                  .
                  .

            A332   Plan and Control Production
             .   A3321 Explode Product
            .   .
            .   .

            A334   Manufacture
                A3341   Make Parts

---

* See Section 7.2.2.

.

.

.

A3345   Stock Parts
A335   Progress Production

.

.

.

A34   Market Products

The data node list is structured similarly, and includes typical statements such as:

D0   RUN COMPANY
   D1   Market Position, Capital, Resources
     D11

.

.

.

   D2   Company Objectives
     D21

.

.

.

   D3   Company Organisation Data
     D31   Company Accounts
     D32   Product Concepts and Requirements Data
     D33   Factory Organisation and Resources
       D331   Production Requirements and Policy
         D3311   Component and Product Design

.

.

.

       D332   Scheduling and Control Systems Data
         D3321   Product Drawing Data and Parts Lists

.

.

.

       D333   Bought-out Items
         D3331   Vendor Information, Price, Delivery

.

.

.

       D334   Process and Job-shop Data
         D3341   Sheet-metal Shop

.

.

.

   D34   Products

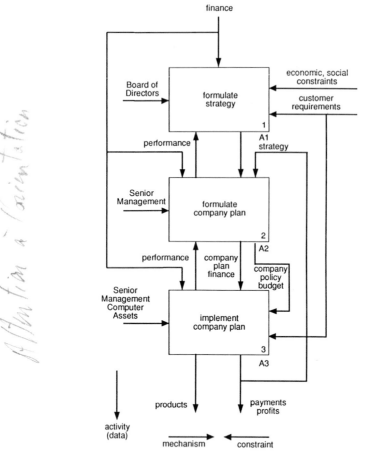

Fig. 2.11   Activity A0, Run Company

The functional and information models representing the A0, A3 and D0 blocks are illustrated in Figs. 2.11—2.13, summarised and amended after [8], using a vertical layout in contrast to the SADT staircase arrangement used to represent dominance, and other slight configuration and notation changes. The revised form highlights the scattering-type and hierarchical concepts inherent in the structured analysis approach.

The A0 Run Company arrangement of Fig. 2.11 overviews and aggregates the operation of the company. The 'finance' input to activities A1, A2 and A3, represents company investment through borrowing and share capital. The 'formulate strategy' activity A1, sets broad company objectives subject to the constraints of external economic and social influences and customer requirements, and the decision mechanisms produced by the board of directors. It

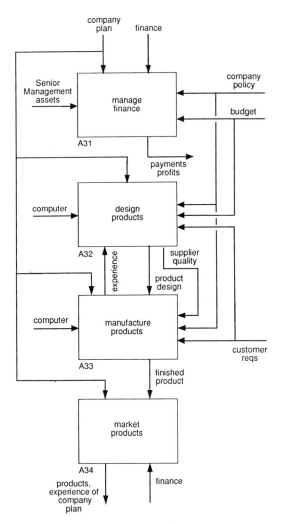

Fig. 2.12   Activity A3, Implement Company Plan

provides the framework required to 'formulate the company plan' activity A2, and ultimately to 'implement the company plan' as activity A3, which operates to achieve planned objectives in terms of product sales, payments and profits.

The expanded form of the A3 activity, 'implement company plan', in Fig. 2.12, illustrates the interaction of finance and the company plan on the aggregate activities concerned with design (A32), manufacture (A33) and marketing (A34). The diagram shows the necessary interconnection between design and manufacture, with the constraints of manufacturing and experience affecting redesign or modification of products. Similar structured models can be formulated for the wide range of other associated lower level activities.

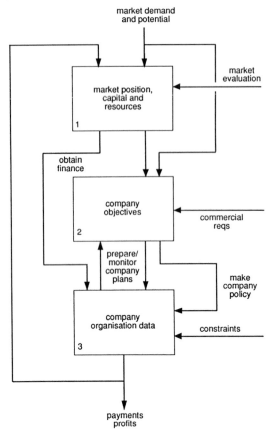

Fig. 2.13   Data D0, Run Company

The D0 Run Company arrangement of Fig. 2.13, similarly overviews and aggregates the information and facilities required to operate the company. The diagram indicates how the corporate strategy is developed, based upon market conditions, available capital and resources. The company objectives lead to the preparation of plans and policies which provide a framework required to operate the company to satisfy the market demand and achieve economic performance.

The structured analysis modelling approach essentially provides static qualitative information relating to the full range of company operations. It does, however, lack a time dimension and means of incorporating detailed mechanisms for designing multilevel–multigoal strategies for overall system control. The technique is particularly useful in structuring large amounts of information which can improve the perception and understanding of the complex mechanisms and information flows involved in integrated manufacturing systems. It can also assist in simplifying and rationalising company organisation and broad based planning and control procedures.

The methods could be further enhanced with the development of interactive computer graphical displays of the interconnected structured models at the various expanded levels of detail, and particularly by possible extensions to include discrete-event simulation which would allow investigation of the effects of planning and control strategies on system performance.

2.5 Manufacturing logistics

Industrial logistics encompass all activities related to the flow of material, components and products, from strategic business planning to design and manufacture, and involves the control of machine processing, material handling, storage etc. Before investment in expensive automation and the development of complex decision making processes is contemplated, it will usually be advisable to consider ways of improving the system logistics and simplifying the manufacturing environment. This can be achieved by controlling product variety and improving factory layout, by changing from a functional grouping of similar machines requiring detailed scheduling to a product grouping or process flow layout. This will lead to reduced inventories and faster changeovers, allowing reduced batch sizes and improved material flow, and give increased productivity and improved customer service, which may be achieved at little extra cost. Accurate forecasting and standardisation are also advisable, with pricing etc. used to force a match between forecasted requirements and customer demand [44].

The current trend in batch production organisation is thus to replace the long established 'process layout' having complex flow pathways handling a large number of parts, with a product type of organisation having simplified flow routes [45]. Process organisation is often characterised by low production efficiency with long throughput times, high work in progress and low rates of stock turnover. It is also difficult to delegate responsibility for due dates and quality to line managers, and this has to be centralised, which is inefficient and costly. With product organisation, decision making and ownership of products can be delegated to departmental groups, which leads to increased job satisfaction and motivation.

Job flow layout in a job shop manufacturing system can be simplified using techniques based on structural modelling and digraph properties [46]. The structural model is obtained by representing the work centres as nodes and the job flow between work centres by weighted directed edges. Concepts of structural controllability and model reduction can then be used to aggregate subsets of variables associated with the original model, and can be related to the changes in system structure advocated by group technology. The structural model reduction of a job shop leads to decentralised control, which is one of the desirable properties of a GT system.

2.5.1 Cell oriented systems
The complexity of the large scale manufacturing system can be reduced and flexibility increased by structuring the system into semi-autonomous cellular product units, operating with local performance measures and linked to other

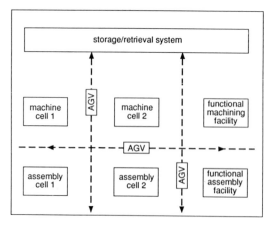

Fig. 2.14   FMS layout example

cells on a contractual basis [47]. The cell concept produces a more interesting working environment which cultivates a sense of belonging, pride of work and competitiveness.

The product cell may include different types of units such as manufacturing cells, warehouse cell, transportation cell etc., with a distributed database supporting local controllers providing local supervision and scheduling functions [48]. An example of a large scale FMS consisting of two machine cells, two assembly cells, functional machining and assembly facilities, an automated storage/retrieval system and automated guided vehicles, is illustrated in Fig. 2.14 (after [49]).

The factory will operate dynamically according to its internal structural mechanisms and decision making processes, and interconnects with and is affected by the dynamics and requirements of the market place. The factory, with its different periodicity of change, must then aim to balance its operations and synchronise in time with the market place, to provide good customer service, minimum inventory, short throughput times, low overheads and high flexibility.

The manufacturing flow problem can be considered as analogous to controlled pipe line flow, and similarly to electrical power flow. Concepts involving minimum power conditions in electrical networks could also be identified, possibly analogous to the requirements to operate production processes at balanced rates just to support the market demand, for optimum use of resources. The principle involving the equal incremental rate loading of electrical generators, using machine cost characteristics and transmission losses to achieve optimum operation [50], might also find useful application in the loading of manufacturing processes.

Unbalanced operation of the production process can produce long throughput times, high inventories, high overheads, and poor customer service, with the

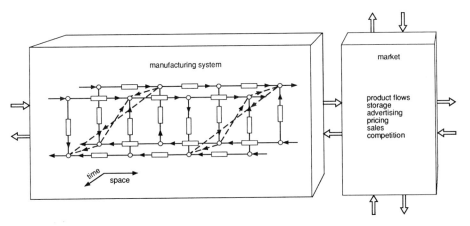

Fig. 2.15 Manufacturing system—market dynamic coupling

resources used unproductively as they switch from crisis to crisis catering for urgent items—possibly similar to the phenomena of chaos in nonlinear dynamic systems. Uniform flow with minimum energy loss and low entropy is desired.

A generic manufacturing system model or manufacturing automaton, coupled to a dynamic market connected to other interacting sources of supply, might also be envisaged for investigating the above phenomena, as depicted in Fig. 2.15. The manufacturing system could be considered as a set of interconnected group transformations within a space–time framework [35] to represent the structured connection of machines, part flows, material handling devices and storage units.

Chapter 3
Advanced manufacturing systems

3.1 Flexible manufacturing systems (FMS)

A flexible manufacturing system is an integrated, computer-controlled complex of NC machine tools linked by an automated material handling system (MHS)—and is essentially an automated job shop [51]. They are designed in the ideal case to process a wide variety of workpieces or part types in medium-sized volumes, on a relatively small number of machines with low manning levels.

3.1.1 FMS types
The different types of FMS can be classified as follows [23]:

 (i) *Flexible machining cell*: This consists of a general-purpose CNC machine tool linked with input–output buffers via an automated material handling system, and is the most flexible type of FMS.
 (ii) *Flexible machining system*: This is a larger configuration of different types of general-purpose machine tools, such as multiple-spindle head changers, with facilities for real-time control of dynamic scheduling and multiple routing of parts. It has the ability to reroute workpieces following machine breakdowns with appropriate grouping of machines.
(iii) *Flexible transfer line*: In this system, each operation for all part types is assigned to one machine, thus producing an ordered fixed route for each part. The system operates like a dedicated transfer line, and consequently is easier to manage and schedule to balance machine workloads. It is, however, less capable of handling breakdowns automatically.
(iv) *Flexible transfer multi-line*: This consists of multiple inter-connected type 3 systems, which increases routing flexibility and the ability to handle breakdowns. Process flexibility is not increased, although scheduling and control are relatively easy, as for type 3, once the system is set up.

The concept of a simple FMS cell with communication links to access data bases containing part programs, machining data, robot control programs, inspection programs etc., and command and monitoring links from and to higher levels, is illustrated schematically in Fig. 3.1 (after [52]).

Overall control will be provided by a cell or master minicomputer, which co-ordinates production at the cell-machine level by scheduling work routing to maximise machine utilisation and production. It also monitors system performance and throughput, and particularly any breakdowns of machine tools and equipment. A DNC computer may also operate at the machine level, to initiate

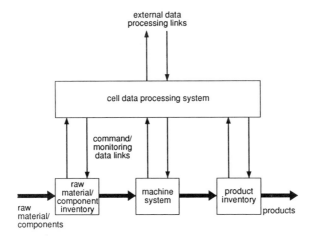

Fig. 3.1 Concept of the FMS cell

and report on the selection and completion of machine tool programs, together with a CNC unit connected to each machine tool at the bottom layer, receiving programs from the DNC and directly controlling the machine tool [5]. The CNC computer will also contain diagnostic programs for detecting and reporting machine tool failures. The generic structure of an FMS complex, incorporating cell scheduling and machine level controllers, is depicted in Fig. 2.6. The strategies and decision making processes required to control FMSs effectively are usually complex and difficult to develop, and will require the support of AI architectures and knowledge based systems.

The large manufacturing complex may incorporate a number of flexible manufacturing systems or cells, the complexity of each depending on the part spectrum and the number and types of machining operations needed to fabricate the parts required. Computer aided design, process planning, production scheduling and control activities may be operated independently in the individual FMSs, or may be integrated into the overall factory plan as envisaged in the concept of computer integrated manufacturing.

3.1.2 Benefits of FMS

The benefits of flexible manufacturing systems are often justified on the basis that a workpiece in conventional batch manufacturing can be idle in storage and waiting in queues for a large proportion of the residence or transit time in the factory—of the order of 95% [10, 25, 42]. The average machine also spends approximately 50% of its time waiting for parts, and in a batch-type shop may be cutting metal only about 15% of the time [53]. On the other hand, metal-cutting times approaching 70% to 100% of the available time have been reported for automated systems [5].

Productivity is thus extremely low from the point of view of machine utilisation in batch manufacturing. This results in high inventories which requires increased working capital to finance the work-in-process [54]. Flexible manufacturing systems can make significant improvements in this respect, and provide a self-contained facility that can automatically plan, schedule, handle material, machine, supervise and inspect production runs. They can have the efficiency and low-cost benefits of a well-balanced transfer line, with the flexibility of the job shop and ability to machine multiple part types simultaneously [55, 54]. They provide the potential for high machine utilisation, shorter lead times, smaller in-process buffer inventories, lower unit costs, high product quality and more rapid response to market demands, although the advantages are often difficult to realise [10, 25a, 23, 56]. The FMS also has the potential for significantly higher capacity than the equivalent flow line system, and by taking advantage of the available diversity in routing they can be less affected by external disturbances such as variability in processing times and machine breakdowns [51].

3.2 Computer integrated manufacturing (CIM)

The concept of flexible manufacturing forms a component of a global strategy—CIM—which is aiming to integrate all company activities into a unified management organisation, using a large scale computer hierarchy. It is expected to embrace corporate planning, marketing, sales, product design, manufacture, planning, scheduling, real-time machine control, material handling, assembly, quality control and dispatch [56].

CIM is an evolving concept and not a technology, and the implementation of the full facility is an ideal which may only be attainable in a new installation. There is, however, a growing demand for progression towards these goals in existing factories, since the potential rewards to be gained from complete integrated computer control are believed to be substantial. They are expected to provide lowest cost production, small or negligible inventories, short product cycles and minimum lead time to market [7]. However, the investment, commitment and risks are significant, although it is generally agreed that the need to attain integrated manufacture cannot be ignored.

A generalised multilevel–multigoal framework, indicating the interconnection of activities and information flows within the overall hierarchy of a CIM system, is indicated in Fig. 3.2. The CIM philosophy is based on a bidirectional, or forward and reverse, flow of information from the physical layer, including manufacturing cells, machine tools, robots and material handling systems, to the highest corporate level involving strategic goals and management control. The information flows between departments in the organisation and the different activities will be integrated with respect to corporate requirements, and data will generally be made available to all monitoring and decision processes. The computer controlled hierarchy will integrate the vertical levels from planning down to operational control, and all activities in the generic case will be linked horizontally [57]. This will require fast and efficient data interchange, using local area networks (LANs) with hardware/software standards and protocols, to link computers, terminals, and a variety of equipment and activities such as machine tools, CAD systems etc.

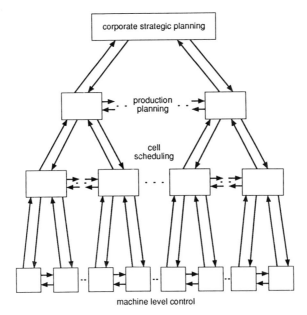

Fig. 3.2   Generalised multilevel–multigoal framework

Computer technology and software support packages presently exist for automating the main activities in CIM, including computer aided design (CAD), computer aided manufacture (CAM), manufacturing resources planning (MRP II), and other facilities including, for example, material handling and inventory control. The ambitious goal of CIM is to integrate and coordinate effectively these presently decoupled and so-called 'islands of automation' with all other supporting business activities. By this means, CIM will view the overall business as an integrated entity and not as a conglomeration of different and separate activities. Each 'island' will then have access to all other activity components, and complete integration will be effected beyond the vertical hierarchical linking arrangement within a more dense bidirectional structure as illustrated schematically in the ladder network/scattering-type configuration of Fig. 3.3.

Fig. 3.3 illustrates the internal signal flow paths which could exist between the external input–output variables in a completely interrelated complex incorporating all possible links. It demonstrates the internal complexity with a limited number of external connections to the 'environment', which is characteristic of large-scale system problems. This enforced internal configuration would probably prove too dense and complex for practical implementation in the manufacturing system, although the formulation of the scattering problem could provide a suitable framework to assist in the analysis and perception of such cellular-type systems, which have previously been considered to be analytically intractable [35].

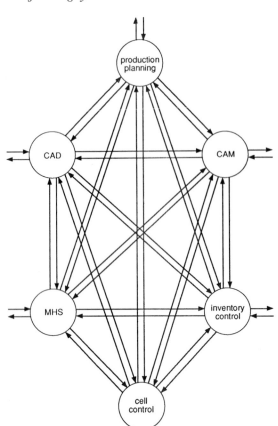

Fig. 3.3   Integration of 'islands' of automation

3.2.1 Problems and progress

The introduction of CIM will require the development of sophisticated AI architectures and the integration of a wide range of complex multilevel–multigoal operations, which will introduce formidable organisational and human problems. The first steps require a detailed investigation of all company practices and manufacturing methods, and it is generally believed that significant benefits can be achieved by first reorganising existing facilities and improving factory layouts, before implementing a scheme of automation [29b]. The market must also be analysed to determine the need for future products and required product variety. Changes in conventional management structures to match the new technology and to achieve the benefits of a new CIM regime—with fewer middle management functions—are also expected [20, 58].

Progression in small steps towards the CIM concept, in conjunction with a flexible long-term company plan, is considered advisable [29b]. A methodology has been proposed for the incremental evolution of the Factory-of-the-

Future—the CIM Open Systems Architecture (CIM-OSA)—in Esprit Project 688 [29c]. This allows the creation of generic CIM structures, including business activities, information flows and communication networks, within which specific applications can be defined and operated, using a set of ideal model building blocks and guidelines. There is, however, no universal model that will fit every case, although there are published examples [59—61]. A functional diagram of a manufacturing enterprise summarised from a Digital Equipment Corporation diagram [20] is shown in Fig. 3.4. More detailed diagrams will be required to identify all activities and information flows.

There are presently few if any suppliers of a complete CIM system, and few companies will be able to afford the large scale investment required. Developments are also presently inhibited by incompatibility problems in linking equipment from different vendors, and by the difficulties of justifying a CIM development economically. Typical audit criteria will need to include assessment of reductions in inventory levels, product lead times, labour costs and improved quality and response to product change [58].

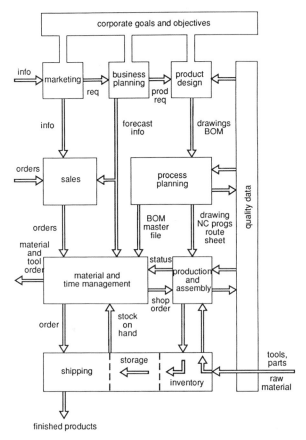

Fig. 3.4 Functional diagram of a manufacturing system

CIM systems will need to operate with large distributed data bases and incorporate facilities for rescheduling and adaptive control, with uncertain conditions resulting, for example, through unpredictable changes in market conditions, and by machine failures and tool wear. Machine cells would also be expected to be capable of self-diagnosis and -repair.

The ultimate concept of CIM, co-ordinating every operation from customer order, through design and manufacture, to delivery, is an extremely difficult and complex objective that might never be fully realised, although some progress is being achieved [7]. The development of general methods for company-specific solutions is required, although detailed generic models of FMS/CIM architectures may not be appropriate across different companies. Also, the CIM approach, which will deal essentially with how the company conducts its business, may not necessarily imply total process automation.

A universal solution to the CIM problem is not available, and present solutions are domain oriented since there is no standard factory. Preliminary work within the Esprit programme has, however, begun on the mapping out of a generic CIM structural model encompassing product design and manufacture [176]. The model will define CIM system functions, the nature of the data to be exchanged, and the constraints and rules governing the exchange. The description presently consists of nine blocks representing product and process development, production planning, materials and production management, equipment control, quality assurance and maintenance. Each block is further subdivided into detailed functions, and data flow is represented graphically in a structured SADT-like style. A function/data correlation matrix is also used to aid in the regrouping of functions into subsystems possessing strong internal and weak external connections.

Chapter 4

Data base systems—Communication standards

4.1 Introduction

A computer integrated manufacturing system involving many multilevel operations will require an extensive distributed data base and comprehensive management information system, containing all data required to design and manufacture a product. The generic form of an information system architecture is illustrated in Fig. 4.1 (after [52]).

The manufacturing system will inevitably generate and need to operate with vast amounts of data relating to functions similar to those in Table 2.1, ranging from marketing and sales to product design, production planning and control, manufacturing, quality control, inventory management and dispatch [10]. The information required to design, manufacture and assemble a wide range of products, which could include many thousands of parts, thus creates an enormous data handling and processing problem. Excessive amounts of redundant data can also be collected, processed and distributed, which introduces the difficult but important problem of contracting data to retain the significant information.

A manufacturing data base will need to interact with numerous management and decision making sources at different levels. It will accept and return information to function modules, concerned for example with computer aided process planning, manufacturing resources planning, computer aided design and manufacturing, NC part programs, etc. A data management system will co-ordinate files connected with works orders, process plans, routings, NC data, robots, inspection data etc. At the shop floor control level, the data base will support the detailed operation of flow lines, cells, machines, automated material handling, tooling, fixtures etc. The design of such data bases, providing multiple interaction and efficient exchange of files between many different functions, and able to maintain integrity and security, presents many technical problems [20].

The information system would normally be hierarchically structured within an interconnected computer network, to support the range of strategic, tactical and operational functions performed at different levels of the organisation. The concept of a control architecture incorporating distributed hierarchical data bases containing all facility planning and control information is illustrated in Fig. 4.2 (after [9]).

The Factory Status (FS) data base is structured vertically to include

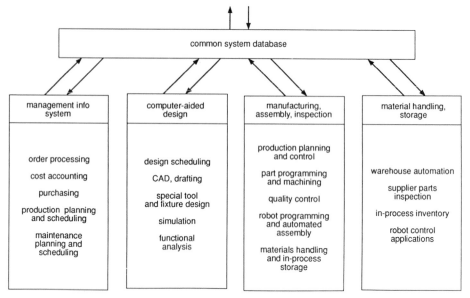

Fig. 4.1   Generic information system architecture

information at the Shop Control (1), Cell Control (2), Workstation (3) and Equipment (4) levels, within a matrix format, containing row, column elements i, j ($i = 1,2,3,4$; $j = 1,2,3$). The data base for Part Data is similarly structured in relation to Process Planning (PP) and Part Design (PD), with i, j elements ($i = 1,2,3,4$; $j = 1,2$). The planning data base contains all data necessary to manufacture the selected part mix, including for example part dimensions and geometry, grip points, tool and material requirements, process plans for routing and scheduling, and tool location files. The control data base section contains status data for all control systems, tools and robots, and management information data used to track order processing. The top level Management Information and Control System Co-ordinator plays a supervisory role, and the hierarchy permits information to be passed between the supervisory modules controlling each manufacturing operation and between the computing modules, via the data bases which serve as common memory.

4.2 Control cycle—State machines

System control can be operated within a network of interconnected routing- and time-dependent state machines or tables, used to specify desired trajectories and set point conditions [9]. The state tables will be processed in real time, with a cycle time interval governed by the nature of the local control action. The cycle will include sampling, table searching to match the measured variables, execution of routines and output generation, and must be short enough to ensure stability and to maintain the system within acceptable levels of performance.

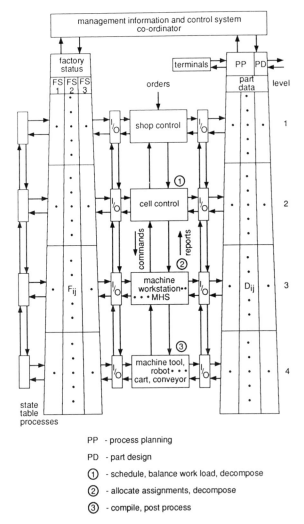

Fig. 4.2 Factory control system hierarchical architecture

The state machines will be integrated into the production management system, to provide the basis for adaptive control at each of the hierarchical levels indicated in Fig. 4.2. The control architecture of Fig. 4.2 indicates the several tiers of inter-level data flows, with each level incorporating intelligent state table decision processes, within a distributed structured data base, used for integrated model-based tracking and decision making [10].

The highest level will co-ordinate group and local factory operations by communicating horizontally with other factories, and vertically with lower-level production activities to provide command information and subsystem goals relating to machine and material requirements. Each level transmits status information back to the higher level and communicates with and supervises lower units to establish operating levels and subgoals. The lowest machine level will implement adaptive digital control of machine tools, part inspection, condition monitoring etc., under changing environmental conditions, using dedicated microprocessors and programmable logic modules. Control algorithms will be generated and compiled at the upper levels using a high-level language, and executable programs transmitted to each manufacturing unit.

Typical features of the hierarchy will include process computers and relatively slow communication bus systems at the upper level, and microcomputers and fast communication at the lower levels. Fig. 4.3 illustrates a conceptual communications framework for connecting subsystems (S_i) via two-way message mail boxes used for transferring commands and status information between levels, within a shared common memory data base. The

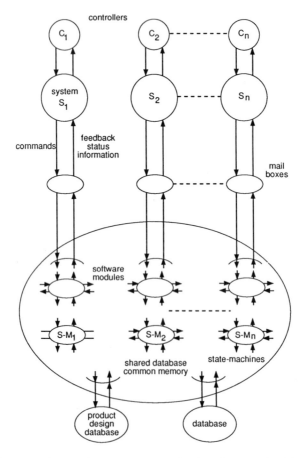

Fig. 4.3 Conceptual control systems communications framework with distributed data base

mail boxes are accessed with commands from single machines, and can provide control and status information, updated every cycle by the state machines, to any other system.

4.3 Communication standards

The effective integration of manufacturing functions as envisaged in the CIM initiative requires common data processing procedures and protocols, and compatibility of computers and communications equipment. Equipment, however, is often incompatible with regard to hardware interfacing and software communication, and few vendors can supply all the needs for integrated automation. In this situation, custom software such as a protocol translator and storage buffer units to provide a common interface at each machine, and modified hardware connections and logic signal levels, will be required to effect communication [9, 10].

The Manufacturing Automation Protocol (MAP), first introduced in 1984 by the General Motors Corporation, was initiated to overcome the problems of incompatible communication systems and to enable equipment at different levels of the hierarchy to share a common language. The MAP standard specifies hardware and software protocols within the structure of an ISO/OSI (International Standards Organisation/Open Systems Interconnect) seven-layer LAN (Local Area Network) reference model developed for computer networking [7, 62].

MAP is designed to link together all production functions, such as order entry, inventory control, robotics, NC machines and quality control. Standard field busses for high-speed real-time communication at the shop floor equipment levels, involving, for example, sensor and actuator data, are also under development [29*d*]. The OSI Reference Model separates the functions needed for network communication into component layers, as indicated in Fig. 4.4 (after [63]). It provides a framework for the structural definition of a communications system, but does not specify the details of the component layers. Each layer connects vertically upwards with adjacent layers, and component protocols are selected to fit into the layers which are suited to the particular problem.

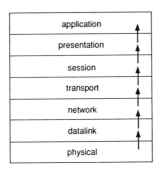

Fig. 4.4 OSI seven-layer reference model

A reduced set of three-layer standard protocols (Proway–Process Data Highway), proposed by the Instrument Society of America (ISA) and based on the OSI reference model, is also under development for real-time communication in small high-speed networks required in process control. The Proway architecture with subsets of the MAP standard forms the basis of a lower-cost, three-layer architecture, called Mini-MAP or Real-time Segment, for high throughput communications between programmable control devices. MAP Enhanced Performance Architecture (EPA) is a standard being developed to provide the same high-level MAP protocols for communication within local manufacturing cells and other time critical applications [64].

A hierarchical model indicating functional groupings and related communications and data processing requirements has been established by the National Bureau of Standards (NBS). The hierarchy and typical networking standards are outlined as follows [62]:

Plant corporate: MAP, TOP, SNA, X25, Ethernet
Area: MAP
Cell: MAP/Mini-MAP
Workstation: Mini-MAP
Equipment: Mini-MAP or proprietary network and 'hard-wired' I/O modules

The hierarchy of protocols has to be defined in terms of code, message and data formats, status words, and the method of error detection and correction. Communication requirements in the hierarchy vary widely—thus the mainframe computer system at the corporate/plant level will be programmed typically for batch data processing for business purposes, and its architecture will not be structured for short-term process control. At the area/cell level, multitasking computers capable of high-speed data processing, will 'front-end' the mainframe computer and co-ordinate work assignment. They will typically download application programs to the machine-level control devices and programmable controllers in relatively short intermittent time periods (seconds). The workstation/equipment programmable controllers and input/output devices provide real-time (milliseconds) scanning and machine control using bit-based data formats.

Graphics capabilities may also be required for data entry, CAD, statistical analysis, process planning, modelling, process simulation, and for the display of general facility status. Advanced fifth generation architectures for the completely automated CIM factory of the future will also incorporate knowledge based expert systems, VLSI technology, parallel processing hardware (such as the transputer) and very high level programming languages.

4.4 Data exchange

Direct translator and neutral format data structures can be used to overcome the incompatibility problem in exchanging data between data bases.
Direct translator: For CADCAM data exchange, the direct software translator will read a specific CADCAM system data base format and convert it to another specific data base format. This requires an 'in' and 'out' translator for

each exchanging system, and software requirements can be extensive for exchange with many other systems, as depicted in Fig. 4.5 (after [14*f*]).
Neutral format: The neutral format data structure is independent of all CADCAM internal formats, and each linking system only requires a post- and pre-processor for conversion between the neutral format and the system's internal data base. The simplified structure is illustrated in Fig. 4.6 (after [14*f*]).

4.4.1 Current standards [14*f*]

(*a*) *IGES*: The Initial Graphic Exchange Specification (IGES) standard was initiated in the late 1970s as a project of the US Air Force ICAM (Integrated Computer-Aided Manufacturing) program [9, 20], and adopted as an ANSI standard in 1980 [14*g*]. It permits the exchange of data between different

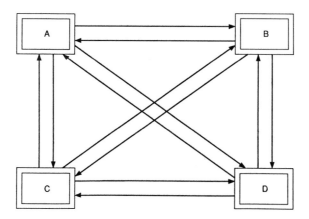

Fig. 4.5   The direct translator

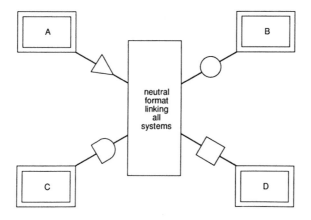

Fig. 4.6   The neutral format

vendor CAD systems, by providing a neutral file format and appropriate pre- and post-processors. It includes facilities for handling a wide range of geometrical constructs including solid geometry and applications such as finite element data. The major drawback is the difficulty in retaining the structure of a CAD drawing, and in transferring design data to the manufacturing data base. The neutral file created can also be very large, requiring excessive processing time.

IGES is, however, the most widely used of the design transfer standards, and has been reasonably successful with two-dimensional drawings and three-dimensional wire frame models. Few full scale implementations of IGES are reported, and new competing standards are being developed as summarised briefly below.

(*b*) *VDA–FS*: Verband des Automobilindustrie Flach en Schnittstelle, is a German national standard, developed by the German Car Manufacturers Association, which covers points and parametric polynomial curves.

(*c*) *SET*: Standard d'Echange et de Transfer, developed by Aerospatiale, is a French national standard. It is a more compact form for data exchange than IGES, and has facilities for transferring geometry, including rational polynomial surfaces and structure. It is being developed to include finite elements, constructive solid geometry and NC tool paths.

(*d*) *GKS*: The Graphics Kernal System is a European graphics standard for transferring 2D graphics via a standard subroutine interface. The associated Computer Graphics Metafile (CGM) format transfers graphics data.

(*e*) *STEP/PDES*: The Standard for Exchange of Product model data and the Product Data Exchange Specification, combine desirable features from other exchange formats, including IGES and ISO.

Chapter 5
Product design—CADCAM

5.1 Introduction

Engineering manufacture is concerned with bringing a product from conception to the market place, and the design process must permeate all aspects of this activity. The objective should be to design a reliable high quality product that the market needs at an acceptable cost, and to maximise customer service.

The primary function of the designer is to balance the design requirements against the constraints of function, materials, manufacturing methods and cost, and good design requires creative and conceptual ability, and the means to convert an idea into information from which a product can be made.

Design plays a vital role throughout the manufacturing process, and decisions and data produced at the design stage have direct effects on other main functions such as production and marketing. It is generally accepted that design must become a more integral part of the CIM process, and take into account the constraints imposed by the specific manufacturing facilities. There must be a more closely linked information flow between computer aided design and manufacture using an integrated product and production process modeller and common data bases, which will allow design changes to be implemented rapidly. The interconnection of the design process with the integrated manufacturing complex is illustrated in Fig. 5.1.

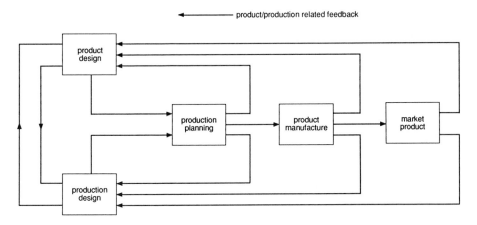

Fig. 5.1 Design for automation

The use of computer aided design tools is changing the design process radically, and the ability to explore alternative solutions and to evaluate design on a 'what-if' basis, and the application of simulation and data base techniques, is having a significant impact, particularly in the manufacturing field. The main thrust of new developments towards enhanced CAD must be directed towards automated design, incorporating a classification of design factors, manufacturing constraints, and integrity and verification of design. Automated techniques must also be developed for directly translating design data into process plans, parts lists and manufacturing instructions.

The introduction of new technologies and advanced manufacturing methods is now changing the nature of the competitive challenge, and if the company is to gain increased market advantage, it must react to the need to manufacture changing products using good innovative design. This will require design engineers with advanced skills in product synthesis and analysis, and able to make decisions under uncertainty and to assess the effects on product design of related manufacturing functions.

5.2 The design process

Designing a product and a system to manufacture the product requires innovation, and is not readily amenable to precise scientific formulation. In broad terms, however, the design process ranging from perception of a market need to product design, can be considered to comprise the following stages [65–68], which represent a step-by-step advance from an initial qualitative clarification stage to a quantitative detailed specification or requirements list based on functional requirements and constraints.

- *Clarification of the task*: Involving the collection of information about requirements and constraints, followed by the generation of detailed specifications
- *Conceptual design*: Providing solution concepts satisfying the product specifications, and relating inputs and outputs through a function structure
- *Embodiment design*: Involving the successive development and refinement of selected schemes leading to a definitive product layout, which provides a best solution with minimum parts count and short assembly times
- *Detail design*: Giving final details and documentation

At the conceptual stage, the design is fluid and unstructured and will be based on creative and innovative thought and 'fuzzy' and amorphous procedures. It can be viewed as a multistage process of subfunctions, verifying and checking a broad range of possible solution sets [14b]. This phase is not presently supported extensively by CAD facilities, although they do have a role in providing means for the creation, manipulation and storage of graphical information, which can be used to translate what may initially be a vague and ill-defined concept into a precise product design.

At the embodiment stage, the design becomes more focused on requirements involving performance specification, shape, material selection, methods of manufacture, testing, maintenance, cost, competitiveness and marketing.

In the detail design phase, the final arrangement, form, dimensions, and material and surface properties are determined, and final detailing with verification of manufacturing feasibility, costs and standards are completed.

The problem can be viewed as an iterative and evolving process, incorporating experimentation, communication and learning, as depicted generically in the scattering-type framework of Fig. 5.2. Fig. 5.3 (after [14a]) illustrates a more detailed systematic structuring of the product design process incorporating inter-stage feedback, tentatively identified by French [67] and developed by Pahl and Beitz [68].

A study of all practical design alternatives is extremely difficult and often impossible because of the large number of choices inherent in the design problem. A combinative procedure has been proposed [67] using a functional analysis and option tables. The tables can be constructed using bases incorporating alternative means of performing functions and different configurations, and using a classification based on known solutions of similar problems.

The design problem needs to be structured systematically into a sequence of trial-and-error decision making processes, and a logical approach used with a measure of intuition and the partial re-use of previous experience and designs, to achieve the best feasible solution. The process will also involve consideration of the following criteria and design factors [69]:

- *Design for economic manufacture*: Involving effects of design on manufacturing methods and production control, and resulting costs of material, labour, machining, tolerances, material handling, fixturing, assembly, stock holding, inspection etc.

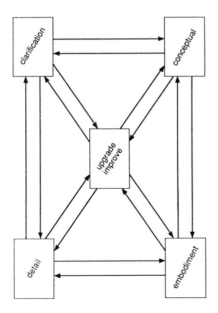

Fig. 5.2 Generic framework for design process

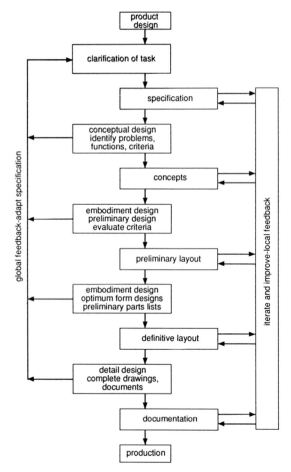

Fig. 5.3 Systematic structuring of the product design process

- *Design for quality*: Based on performance and test specifications, control of quality through design
- *Design for performance*: Incorporating functional requirements, performance testing, design margins on tolerance
- *Design for life-cycle costing*: Design for maintainability, product servicing modules, design for optimum product life and reliability
- *Design for safety*: Safety hazards, failure survival and intrinsic safety
- *Standards*: Consideration of a range of standards, BSI etc.
- *Marketing*: Role of marketing in engineering design

5.2.1 Structural problems
The development of automated techniques in manufacturing will require means for quantifying and integrating the above form and function factors into the design process, using empirically based modelling techniques and knowledge

based systems. This will require the modelling of multidimensional design characteristics, which introduces complex problems and questions such as:

- Can the design problem be quantified in terms of an identified model and an objective function, with the design process considered as a systems problem involving feedback and iterative loop decision making?
- How do we establish structural relationships linking the conflicting requirements of design?
- How do we incorporate 'good' characteristics related say to past designs?
- What characteristics make a good design?
- Is it feasible to produce a structured multidimensional data set relating to design?
- How can this information and knowledge be related to a scalar objective function defining 'good' or 'bad'?

It is interesting to conjecture that we are involved in some of these issues when decisions are based on hunches and acquired knowledge relating to previous good designs, and possibly obtained using stored imagery and subconscious multistage decision making.

Representation of the design problem, using say a multidimensional data set with empirical structural relationships to interconnect the interacting factors involved, would permit the development of a sequential numerical procedure based possibly on group handling or multilinear algebraic techniques for parameter fitting and model building, using a scalar objective function to define a 'best' overall design solution. An automated design process incorporating a design model-based objective function could proceed as a sequential interactive tuning procedure such as illustrated in the structured form of Fig. 5.4. The approach would effectively impart an empirical structure on the design process, and provide decision making support which could be integrated with the processes involving hunches and intangibles to provide a final design solution. Other approaches have considered the design process viewed in terms of logical, causal relationships between inputs and outputs, states and environmental factors, associated with the transfer of energy, material and signals [14*b*].

Existing CAD software will need to be extended beyond the existing facilities available for finite element analysis, to incorporate routines for model testing and performance evaluation, including the study of physical properties such as deflection, pressure and temperature profiles, vibration and the dynamic behaviour of engineering components and systems. Relating function to shape is difficult, particularly for a specified response and dynamic behaviour, and little is known in general terms on how to integrate the various design factors with functional requirements. These developments will also require enhanced graphical data processing with parallel architecture, to obtain improved responses with large data bases. The integration of research on the engineering properties and applications of new materials such as ceramics, glasses, metals and polymers and the growing range of composite materials, and their impact on the design process, would also be desirable.

A quantitative approach has been used for the selection of material properties in the design process, with relative costs considered to be governed by design, manufacturing and material stock, in the form indicated in Fig. 5.5 (after [71]).

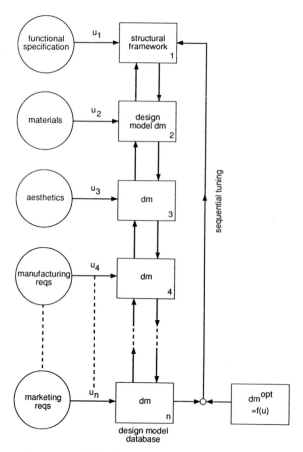

Fig. 5.4   Interactive sequential design process

An interactive numerical procedure is used to quantify the relative importance of metal properties, and the user has access to product design, stock position, the physical properties and applications of metals, etc. Scaling factors are assigned to each material property to quantify its ranking of importance in the selection process, and a method involving weighted property characteristics and a polygon of design factors and material properties, which is compared to the size and shape of an ideal polygon of factors, is used to quantify the material selection problem.

5.2.2 Taguchi approach to parameter design
The Taguchi methodology for optimising product and process designs, developed in Japan by Dr Genichi Taguchi, provides an important quantitative systematic approach which can lead to improved products. The philosophy essentially includes three steps: (i) system design, (ii) parameter design, and (iii) tolerance design [72, 73].

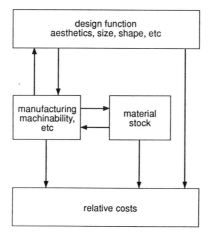

Fig. 5.5 Interactions between design, manufacturing processes, raw materials and costs

Parameter design is the most crucial step for achieving high quality, and requires the determination of product parameters or process factor levels which optimise the product's functional characteristic with minimal sensitivity to 'noise'. It involves data analysis using orthogonal arrays, signal-to-noise ratios and a quadratic loss function related to the deviation of the product's functional characteristic from its desired target value.

The functional characteristics are affected by controllable factors such as choice of material, cycle time etc., and noise (or uncontrollable) factors which may be difficult or expensive to control, such as manufacturing imperfections, process factor tolerances and the effects of machinery wear. The aim is to select the controllable factors such that the product (or process) is least sensitive to changes in the noise factors.

The controllable factors are tested for significance, robustness and cost reduction over specified levels, using an experimental orthogonal design to obtain the best combination of product parameter values and operating levels of process factors which are least sensitive to change in environmental conditions and other noise factors. Tolerance design can then be employed if the reduced variation obtained through parameter design is not acceptable. This requires reduction of tolerances on product parameters or process factors, which involves expenditure for better grade materials, components or machinery.

Consideration needs to be given to the effects of interaction between the controllable and noise factors. This is the key to achieving robustness against noise, by changing the noise factors in a balanced fashion during experimentation, and allows preferred parameter values to be determined by the analysis of an appropriate signal-to-noise ratio.

In practice, a number of controllable factors that are least influenced by changes in the noise factors, and which could affect the desired performance parameter of an assembly, are identified [72]. Each controllable factor f_{ci}^{j} ($i = 1, \ldots 4$, say), is quantised over levels j (say 1, 2, 3) and tested at each of the noise levels f_{ni}^j ($i = 1, 2, 3; j = 1, 2$ (say)). The resulting information is used to construct matrix arrays of experimental data associated with the effects of the controllable factors and the noise factors set at different levels during each experiment. These arrays, together with columns representing estimates of interaction effects between the controllable factor levels and noise, are used to form an aggregated matrix array representing the complete parameter design layout. Study of the response variations provides a combination of controllable factor levels that could give improved performance and robustness against noise.

Levels of the controllable factors are selected on the basis of the highest signal-to-noise levels $(S/N(dB) = f(y_1, \ldots, y_n))$, where $y_1, \ldots, y_n$ represent n observations of the controllable factors. A statistical analysis of variance, using F tests to determine which factors are statistically significant, can be used to separate out the factor effects in terms of the S/N ratio and mean response.

The method has application to product design, and provides a choice which leads to the most desired and consistent functional characteristics. Major noise factors considered during the early stages of the product design cycle could include external causes such as temperature, humidity or power voltage variations, and product deterioration such as wear and tear of materials or component parts [73].

5.3 Product design data base

The following data structures are often used in dealing with engineering data:

- *Tabular*: as in supplier catalogues of components and systems
- *Tree structures*: relating to the connection of components and system subassemblies
- *Mesh structures*: consisting of a graphical connection of node points, as in piping and wiring diagrams or the geometrical boundaries of a shape.

The usual type of commercial data base is not designed to handle these structures, and their accommodation is not straightforward. New or modified data bases are needed which can store and handle the large amounts of graphical and textual data required in CAD, typically amounting to the order of 100K A4 pages of data and diagrams for a small design office [14c]. The storage of this amount and type of information is a formidable task, and is likely to require special facilities compared to those available in a commercial data base. The storage of dense graphical structures, particularly, requires large computer stores, which severely restricts the number of such graphs which could be stored, say on desk top systems. The problem requires study of the means of compressing large frames of data while retaining the significant or essential features of a graphical design in order to economise on storage.

The part numbering of components using standardised codes to identify functional properties, as used with standardised frame sizes for electric motors and bearing types and sizes, has been proposed to facilitate the accessible storage of component data [14c], and could have important applications in CAD data base design. The ability to incorporate design calculations in the data base with interactive manipulation of equations in an algebraic format, and the development of so-called Parametric Programs and Plastic Modellers able to produce drawings of components with variable dimension and loosely defined shapes, and possibly interacting with the design of other components in the graphical data base, would also extend considerably the automation of the design process.

CAD systems will usually be unable to assist during the conceptualisation stage of design, although the availability of specialised data bases, incorporating catalogue-type video disc information or links to national standards and component information data bases, could prove extremely useful [14a]. Effective application of CAD technology is reliant on a central product data base produced to a consistent and coherent format, and accessible by other design and manufacturing activities. This would increase overall design effectiveness, particularly for established product designs based on product families or standard ranges.

New ideas and concepts are needed for developing a comprehensive product design data base, able to store and search for combined parts lists and graphical information, and integrated with manufacturing. Mathematical symbols and simple algebraic expressions, similar to those available in the MACSYMA computer algebra program, are also needed to support analysis in the design process.

A computerised planning system will require implementation of a formal structure and data base to enable design information to be transformed into production design [3]. Most objects can be modelled using a set of primitives such as a polyhedron with n faces each of which can be modelled using an ordered list of the vertices (points) or edges (lines). Curved surfaces will require the use of codes to distinguish them from polyhedron surfaces. Tree structures linking faces, edges and vertices can be used to represent the topology and basic geometrical relationships of an object which are required to manipulate and display the object. Additional technological information, such as tolerances, dimensions, colour, manufacturing constraints etc., will also be required for a complete representation, which can be stored in a list structure with the element geometry connected using data pointers.

CAD solid models are often developed using a combination of primitives such as blocks, cylinders and cones to form more complex shapes, and the Boolean operations of intersection, union and difference [74e]. Binary relationships can also be used to describe the component and other added factors, such as the blank, machines to be used, tools and tolerances, to form the knowledge base system [24e].

5.4 Intelligent knowledge based design

The design process is concerned with creating products based on criteria and requirements as discussed in previous sections. This requires a wide breadth of knowledge of the factors relating, particularly, to materials, surface finish etc., and of the design features affecting the manufacturing and assembly operations. It is often difficult if not impossible to formulate the problem precisely or analytically, and knowledge based systems are being developed to support the design process, particularly where expert competence, experience and judgment play an important role. Knowledge based systems are also well suited to problem solving where symbolic representations and pattern matching are equally as important as numerical or algorithmic procedures [75, 76a]. They allow the incremental development of knowledge and rules, and the expectation is that they can build up experience, which is not lost to succeeding generations of designers.

The output of an intelligent knowledge based design system would be expected to provide a complete description of the product required for the manufacturing stage, together with formal representations for [18e]:

- Geometric description of product parts
- Material properties, surface finish, tolerances
- Functional description—performance, constraints
- Process plans—materials required
- Manufacturing constraints—materials handling, machining, assembly operations
- Operational plans—tools required, part programs, setting data, cutter paths, operation times
- Maintenance knowledge
- Costing information

The structure of a possible decision making system forming an integrated Design-Manufacturing data base is outlined in Fig. 5.6.

Application areas for expert systems combined with geometric modelling have been identified as [76b]:

- *User interfaces*: With the facility to input existing drawings into the modelling data base using video camera interpretation of 2D drawings
- *Design/manufacturing*: Integration of a geometric parts model data base, defining tolerance, surface finish etc., with an expert system data base containing process planning and NC programming knowledge
- *Analysis*: Feature recognition with analysis of stress concentrations using appropriate finite element mesh generators.

A systems engineering approach to engineering design, involving the integration of geometric form and functional design, has been considered using a rule-based logic programming expert system shell as a framework for knowledge representation and problem solving [14b]. General form features include directed and rotational surfaces and axes of rotation, points and surface areas of application, lines of action etc. Geometrical and topological relations may then be defined between these objects by means of production rules and constraints. The overall system function can be decomposed into input–output sub-function relationships to facilitate a structured approach to system synthesis, using a tensor symbol representation.

A knowledge based numerical design procedure (DESIGNER) incorporating a model based on explicit design characteristics, e.g. length, dead-weight etc., has been developed. The qualitative knowledge is contained in the form of a network or digraph, with nodes representing characteristics and the arcs relationships between the nodes [76c]. Fig. 5.7 illustrates the form of a network

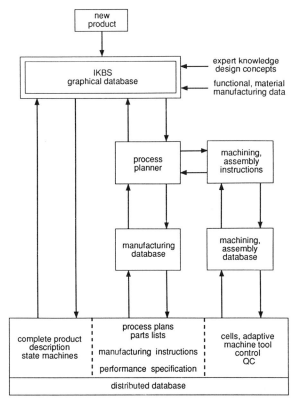

Fig. 5.6 Design–manufacturing data base

representation for a particular application to ship design. A multilevel representation of the functional hierarchy for components in an engineering system has also been used to provide a basic structure of knowledge relating to geometry and function. An example of a simplified functional 'family tree' is shown in Fig. 5.8 (after [77]).

The design of products that are easily manufactured and assembled is a difficult task, and significant reductions in manufacturing costs and increased productivity can result from simple design modifications. Knowledge based expert systems can assist in this respect, by alerting the designer to difficulties that the design presents for automated handling during manufacture, and providing for necessary design changes. A knowledge based system to indicate possible difficulties in component handling for automated assembly is reported [76a], in which the system attempts to model the sequence of deductive stages used in reasoning to solve the component handling problem. Further developments are required, however (as discussed in Chapter 1).

An interesting approach to the problem of Design for Economic Manufacture using a logic-based Prolog expert system, in which data can be supplied

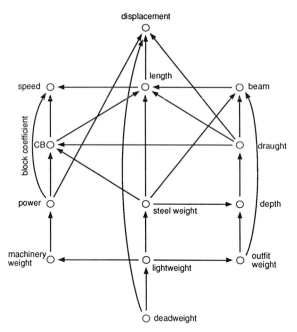

Fig. 5.7 Digraph used for preliminary design

to the shell as textural dialogue or by direct input from drawings, has been reported [74*h*].

Further discussion of knowledge based systems and their applications in manufacturing processes is included in Section 7.2.10.

5.5 CADCAM

The design of a product not only determines its performance and customer appeal, but also affects significantly the manufacturing process. Careful consideration must thus be given to component design, and particularly to the effects of shape and material on methods of manufacture and assembly which will directly affect lead times and production costs.

It is reported that up to 70% of the cost of a product can be incurred during the design phase, and 20% during actual production, although the product design process takes little account of production in many companies [75]. Very large savings in design/manufacturing/assembly costs are thus possible if correct decisions are taken at the design stage.

An important aim of CADCAM is to reduce the design and production cycle time, which can only be achieved by a more closely linked information flow between CAD and CAM. This requires the development of an integrated product/production process modeller, and account must be taken within the design process of the downstream factors and constraints imposed by the

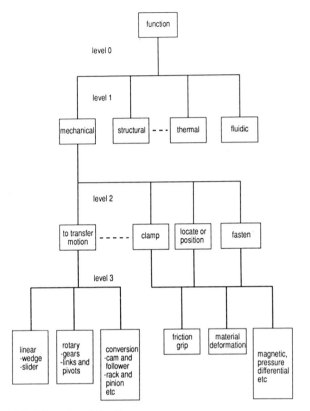

Fig. 5.8 Simplified functional family tree

specific manufacturing facilities and assembly processes involved. This requires a definition of the design rules and simulation of the downstream production processes at the design stage to improve the probability of a 'first time' correct design and the suitability of the design for automated manufacture. It is thus important to consider the integration of product design, process planning and production, although design for production and design for function can often produce conflicting requirements which can only be evaluated in cost terms. The benefits also are difficult to quantify since there are few indices to monitor.

The geometric model created in the design process within a common data base can in principle be used to produce detailed drawings, and then transferred to the manufacturing section to assist in the planning of the manufacturing facilities, including the production of jig and tool designs and NC machine tool programs. The common data base for design and manufacturing will facilitate the rapid implementation of design changes and thereby reduce wastage and redundant data in the manufacturing data base. It will also form a basis for developments towards full computer integrated manufacturing. Developments are leading to the specification of centralised relational data bases that will store data on all aspects of design and manufacture [14d]. The

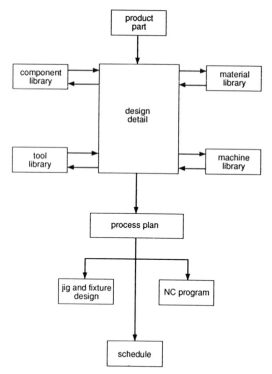

Fig. 5.9 Integrated design and manufacture

aim is to use a single unified system, which will provide rapid iterations of designs through engineering and manufacture.

In future CIM systems, the design activity will be linked directly with the manufacturing data base and automated techniques will convert the geometric definition of a product, obtained using an iterative design process supported with expert knowledge, directly into performance specifications, and initiate detailed manufacturing instructions, including process plans, bill of materials, parts lists, production scheduling and material movement [18]. The majority of present-day CAD systems, however, only prepare drawings or produce CNC machine control data through a manual interface. Current research aims at the automation of the missing link between CAD and CAM, and the automatic generation of process plans from a solid geometric model.

Process planning is the link between design and manufacturing, and an automated process planning system linked to design may involve the specification of part geometry in terms of available tooling. An interactive design procedure could automatically specify tooling and geometry simultaneously, and component material and machines to be used. Fig. 5.9 (after [74a]) illustrates the concepts of an integrated design-manufacturing process.

Chapter 6
Planning, scheduling and control

6.1 Introduction

The technological and economic processes involved in converting raw material to marketable products, and the problems involved in ensuring that the company is viable and using its resources effectively and profitably, are complex and require detailed consideration of the needs for strategic planning at the company level, and operational planning, scheduling and control at all levels of the organisation.

Strategic planning is relatively difficult because modelling and solution procedures are generally unknown, and decision making will usually need to rely on intuition and heuristics. Operational design and planning will need to select and design for future courses of action from among many alternatives, and control action is required to measure and achieve desired operational performance in accordance with higher level objectives and plans [78].

6.2 Strategic planning

Effective operation of the company in highly competitive markets will require a long-term Strategic Business Plan for aggregate requirements, focused on profits and growth. The essential elements in such a plan will be [79]:

- Business Strategy, which is required to define the aims and objectives of the company, and policies for marketing, research and development, production, and for personnel and financial management
- Product Strategy, based on forecasts of long term market demand and developments in product and manufacturing process technology [80]
- Manufacturing Strategy, based on engineering knowledge concerning product design and manufacturing.

These strategies are closely interconnected, and the Manufacturing Strategy, particularly, will be governed by constraints and goals set by the higher level Business and Product Strategies. The central function of the Manufacturing Strategy in relation to the higher- and lower-level activities is illustrated in Fig. 6.1 (after [79]).

The design of an automated manufacturing system, incorporating interconnected computer systems and a wide range of equipment, must be planned on a corporate-wide basis to ensure the formation of effective management structures and the efficient operation of the overall complex. The key to corporate success

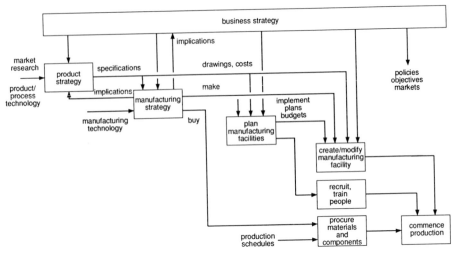

Fig. 6.1   The manufacturing strategy

requires the optimum use of all available resources, and in the final analysis it is the quality of the managers that decides how effectively the activities of the company are organised to achieve the overall objectives.

The development and implementation of the overall Strategic Plan will require selection of appropriate business units and planning teams, to consider critical issues such as:

- Business mission, present practices and weaknesses, current position and performance, required data
- Information flows, effective communications for monitoring, control and accountability
- Market trends, competition
- Investment, strategic options, products/markets
- Action plans, schedules
- Effect of the product life cycle on marketing, production and research and development
- Required planning horizons considering the rapidly changing technology and environment
- How to measure validity and effectiveness of strategic objectives and programme plans
- Strategic guidelines for the overall co-ordination of marketing, production and research and development.

The corporate objectives will be founded on a broad basis of profit, and can be structured heuristically into a hierarchy of sub-objectives or goals including, for example, long term plans, investment policies, marketing activity, environmental issues etc. These then produce constraints and subgoals for the lower level processes involving product design, production planning and control, and sales. The total operation of the company thus involves, in broad terms, the co-

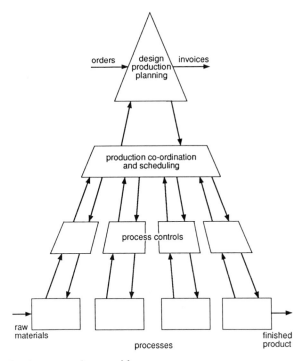

Fig. 6.2 Production control pyramid

ordination of a complex hierarchy of controls, operating on different types of information which becomes less precise and limited at the higher levels. Such integration, involving multilevel decision making and adaptive control of nonlinear dynamic processes subject to stochastic disturbances, represents a complex, large scale systems problem which in general cannot be formulated and solved analytically [81, 78].

A knowledge based model incorporating 'what-if' analysis could be used to give management support, to aid strategic resource planning decisions with long term horizons [29e]. This could include the effect of changes in important factors such as the product market strategy, the impact of technology programmes, international competition, as well as corporate and manufacturing goals. Product demand forecasting also plays an important role in operational planning and control, and time series analysis techniques for spectral analysis and two-dimensional prediction, such as those based on the Karhunen–Loevre expansion [82, 83], could have important applications in manufacturing.

6.3 Production planning and control

The concepts of overall Planning are similar in many industries, and can encompass Design, Operational Planning, Scheduling and Control [84]. The functions form a natural hierarchy as indicated in Fig. 6.2 (after [15]), and link

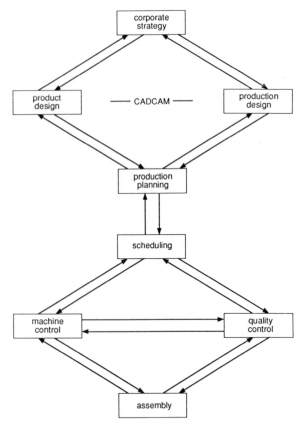

Fig. 6.3 Structured sequence of manufacturing activities

with other main components involving corporate strategy, product design, quality control, assembly etc., in the broad form illustrated in Fig. 6.3.

Each of the subproblems operate according to different local objective functions, and provide constraints on the lower level problems in the hierarchy. During the design stage, appropriate goals may be to minimise forecasted operating costs, maximise system flexibility or to satisfy customer requirements. At the planning and scheduling stages, appropriate objectives would be to maximise production or system utilisation, to minimise inventory or lead times. The control stage may need to operate with objectives relating to quality, system errors and reliability. Throughput will be determined by the mix of jobs, levels of inventory, and the way work is released and scheduled through machines by the setting of priorities.

6.3.1 Production design

Production design is concerned with determining the numbers and capacities of machine tools, buffers and material handling systems, and the required plant layout to match desired production volume and product mix in the face of uncertainties such as customer demand and breakdowns. At this stage, attention must be directed to industrial logistics, that is, to the activities related to product flow, which will be governed by the strategic business plan, manufacturing processes and capacities [44, 85]. The logistics will affect inventory levels, factory and customer delivery performance, and consequently cash flow and asset turnover (see Section 2.5).

The design of a layout configuration for a new or redesigned integrated manufacturing facility is a complex problem, and must be based on a detailed specification of manufacturing requirements, and will require a considerable amount of trial-and-error design and iteration before the 'best' solution is obtained. Typical decisions, such as those required for an FMS design [84] in the medium term, would be concerned with:

- Identification of the range of part types and subsets to be manufactured and assembled—specification of capacity and functional requirements such as machining times, part routing and cutting tools required to meet expected production requirements; material selection
- Number and types of machine tools, tool handling requirements, robots, gripper points, material handling systems, buffers, pallets and fixtures
- Specifications for computer control hierarchy and strategies, and communication between different levels, linking cells, machines and material handling equipment
- Details of operational planning, scheduling and control strategies and objectives, to check feasibility of physical design, operational policies and the effects of breakdowns and bottlenecks on system throughput, queueing and waiting time
- Specification of software development for scheduling, tool management, inventory control, robot operation etc.

6.3.1.1 Buffer storage. Buffers tend to decouple adjacent machines and reduce the effects of machine failures propagating throughout the system. They allow upstream and downstream machines to continue operating provided there is space and material, respectively, in intervening buffers [86]. In practice, when a machine fails, neighbouring work centres will be informed to limit production accordingly, and routing and scheduling policies can be reconfigured to maintain production and reduce in-process inventories.

Increasing storage capacity will increase overall machine utilisation asymptotically towards the efficiency of the least efficient machine in isolation. Increasing the efficiency of an individual machine has the effect of increasing the production rate of the transfer line, and buffer storage contributes most to system production rate when the line is balanced, with similar efficient machines. Small increased amounts of buffer storage can sometimes improve the production rate as much as increasing the isolated efficiency of machines, which may involve considerable capital expenditure.

Analysis of the economics of installing additional buffer storage and other facilities to improve system reliability and efficiency, with the system subject to random disturbances, must consider the best trade off between installation and inventory costs, and particularly the direct effects of storage size on efficiency. The effects of storage and machine failures on overall system behaviour are in general difficult to predict, and theoretical analysis has been restricted to special classes of systems such as the flow shop or transfer line, with two or three machines, as depicted in Fig. 6.4 (after [86]). Difficulties arise with the analysis of larger systems because of the rapid growth of the state space.

Simulation methods have thus been used for machine-buffer analysis, although these are time consuming and provide only statistical information, and are unsuitable as a detailed design tool. Simplified approximate analytical methods are required, and possibilities may exist for the application of scattering techniques [35] using the basic configuration of Fig. 6.4. Other methods have been proposed, based on the aggregation of adjacent machines and buffers into a single machine to produce a reduced two- and three-machine system representation which could be analysed using exact methods [86].

In early methods of inventory control, replenishments were ordered when inventories fell below certain (reorder) levels, without regard to the effects on production which could fluctuate significantly and result in severe bottleneck problems. This would cause lead times to vary widely, and extend forecast demand horizons possibly into periods of financial uncertainty [87]. A more proactive approach to inventory control is now required, providing fast response and reduced lead times, as in the MRP-based procedures which started to be adopted in the 1970s.

Inventories have usually been considered necessary to provide a buffer against uncertainty, although analysis of a simple failure-prone manufacturing system producing a single commodity has shown that the argument does not hold, and that there are ranges of parameter values for which a zero-inventory policy can be optimal even in the presence of uncertainty [88]. This is an interesting result which supports the present tendency towards reduced inventory operations which enforce a more rigid discipline on the manufacturing system. Further work is required to extend this type of analysis to more complex models of multi-machine, multi-product FMSs, in order to investigate whether zero-inventory policies remain optimal for particular ranges of 'mean times between failures' and 'mean repair times'.

6.3.2 Production planning
Production planning aims to organise the available resources in order to operate the business profitably and satisfy customer demand. The function will usually involve a relatively large number of products and processes, and it will be necessary to group equipment and to batch orders and items in accordance with production constraints, and to set up the process equipment to satisfy the working schedule and sequencing of the production flow.

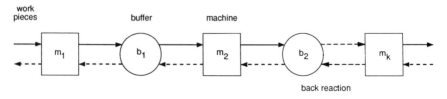

Fig. 6.4   Transfer line with interconnecting buffers

The planning of resources and activities must be undertaken before the system commences operation, and the resulting plan will contain information concerning routes, machines, process parameters, tools and fixtures required. The planning operation will usually be based on production forecasts using past market demand and sales data, and the processing of different lots will require detailed and sophisticated planning models.

The important decision problems involved in aggregate production planning are concerned with determining [28, 3]:

- *Optimal product mix*: Selecting lists of part types for production orders and subsets for simultaneous manufacture
- *Machine allocation*: Partitioning machines into groups and pooling identically tooled machines to improve reliability and performance; allocating pallets and fixtures to selected part types
- *Requirements analysis*: Determining product quantity to be produced in a specified period subject to demand, due dates, plant capacity and operational constraints, possibly using mathematical programming techniques
- *Production smoothing*: To balance the utilisation of production facilities, and adapt to market demand fluctuations

The following detailed planning functions can be identified [26b, 89]:

- *Batching*: Partitioning the weekly plan into a list of parts for production within the next shift
- *Routing*: Defines the process plan or routes for each part type
- *Dispatching*: Defines the sequence of parts into the FMS, subject to priorities
- *Sequencing*: Defines the sequence or rules in which machining, tooling, and transport operations will be performed.

The process plan converts design data into the work instructions required to manufacture and assemble the product [20]. It determines how an order will be routed through the shop, and how machines and tools will be scheduled and sequenced, and gives detailed instructions for machine operations and planned manufacturing times. The activities and decisions required in process planning are indicated in Fig. 6.5 (after [74b]), and the problem can be described in terms of information flows, using a Structured Analysis Design Technique (SADT), as illustrated in Fig. 6.6 (after [90]).

The existence of multiple part types, parallel machines and alternative routings in a FMS, and the relatively large number of conflicting goals and constraints, increases significantly the number of alternative decisions and the complexity of the decision making processes. A function approach to these problems will involve many thousands of binary variables and constraints for a relatively small number of workstations and parts, and considerable computational effort will be required [89]. Production planning over a long period is also generally unrealistic because of system failures such as tool breakages, machine breakdowns and malfunctioning of actuators. Continued updating over a shorter time horizon is thus required when the current plan becomes inapplicable [20a]. Increasing flexibility inevitably increases the number of possible states and dynamic perturbations, which increases the complexity of the control

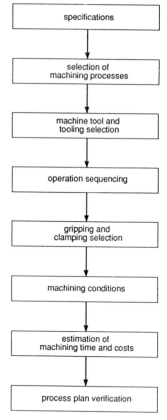

Fig. 6.5   Process planning activities

system. If the plan then becomes inapplicable, flexibility must be reduced because of the limitations of the control system.

Gantt charts have been used traditionally for operational planning and resource allocation, although the manual approach with a large number of conflicting constraints can produce excessive in-process inventories, low machine utilisation and late completion of orders [89]. The trend in production planning will be to incorporate an expert system and model-based heuristic scheduling rules to balance load and to cater for capacity constraints and uncertainty. The future aim will be to generate a process plan automatically using a product design description and a manufacturing knowledge data base.

6.3.2.1 The planning dilemma.   The planning process is affected significantly by the relationships between product forecasts, customer orders, inventory levels, customer service and bills of material. Bills of material define the product and how it is assembled, and are used for the procurement of materials and parts and in scheduling. They provide a basic framework for planning and

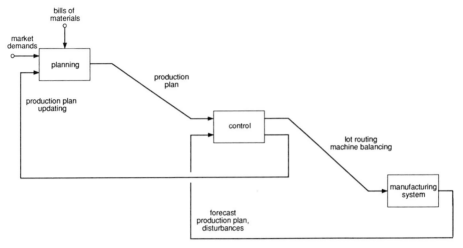

Fig. 6.6 SADT representation of production planning activities

scheduling almost all products, and their integration with design and forecasting directly affects inventory levels and customer service [44].

The bill of material starts with the creation of a master schedule (MPS), specifying part production over time based on forecasts and customer orders, which generates detailed schedules subject to capacity constraints and financial objectives. The master schedule is then exploded, to allow planned start dates and quantities of all items and sub-assemblies to be defined, taking account of stock and lead times. This technique is called material requirements planning, and its objective is to schedule production and procure materials at the correct time and in accordance with the master schedule.

The final schedules leading to the finished products are thus determined directly by material requirements planning and inventory levels and in turn by the master schedule which results from the long range plan based on forecast information and firm orders. This situation can result in a planning dilemma, produced by an 'input–output' mismatch, as indicated schematically in Fig. 6.7 (after [44]).

6.3.2.2. Computer aided process planning. A computer aided process planning system aims to read in design specifications, and use stored knowledge to plan the required manufacturing processes, and is a prerequisite for bridging the functions of CAD and CAM [3]. In general, the input will be a three-dimensional CAD model, with information on shape, tolerances and special features. Time estimates, resource requirements, and material handling and machining part programs, could also be provided for scheduling and control.

Recent developments in knowledge based computer aided production planning have used both variant and generative approaches.

Variant process planning requires a unique data base of component families and standard plans for similar components, and contains no process knowledge *per*

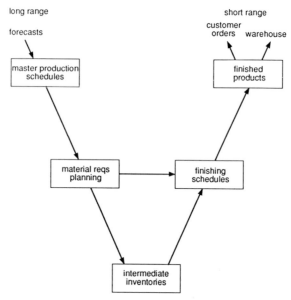

Fig. 6.7   The planning dilemma

se. It provides a sophisticated and reliable planning system although it requires time-consuming generation of data files and is costly to implement. The form of representation typically codes the part to enable the nearest standard process plan to be recalled [74*d*]. A variant process planning system presents problems if there is significant product variation and few similar components, and considerable effort could be involved in adding new families and standard plans.

Generative process planning is more complex and aims to generate process plans automatically for new components, by mating a description of component features and geometry to methods of manufacture and available equipment [74*d*]. It has the potential to automate the production of planning functions such as machine and tool selection, and optimum process scheduling and control, although this requires the identification and transfer of component data and process planning decision rules into a computer-readable format which is a difficult task. The development of generative process planning is complicated, particularly by the difficulties in quantifying process capabilities such as tool life, which has delayed the introduction of a true generative system. Specialised developments have included applications to drilling [91], turning [76*d*], and milling [92], although its widespread use incorporating links to CADCAM and data management systems has been limited.

A merging of the variant and generative approaches is required, as in the SIPPS (Synthetic, Interactive Process Planning System) linked to CAD [24*d*]. The system uses initial symbolic shape descriptions which lead on to a geometric model and detailed engineering drawings, NC part programming and general

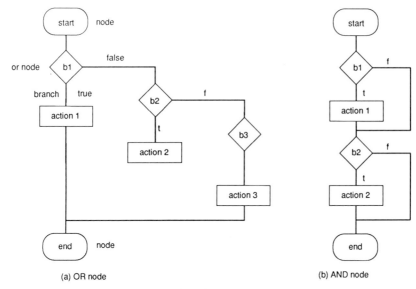

Fig. 6.8   Structured decision tree flowchart

process planning. Variant techniques are used through a flexible manufacturing table using an embedded coding and classification system to group the final codes into categories, and the system can proceed automatically to generate process plans based on component specifications when the table is finalised and all parameters selected.

6.3.2.2.1 Decision logic.   In implementing the concepts of CADCAM, system decision logic must be able to match the process capabilities, described say by 'IF . . THEN . .' expressions, with the design specifications. The decision logic structure of process planning can be described by:

(i) decision trees, (ii) decision tables, (iii) artificial intelligence

(i) *Decision trees* can provide a natural representation for process information. A typical configuration, incorporating an OR node with conditions (IF) and decision statements (true, false) set on tree branches (representing rotational components, holes, tolerances etc.) and predetermined actions (drilling etc.) at the junctions of each branch activated by a true condition, is illustrated in Fig. 6.8, (*a*) incorporating an OR node B1 and (*b*) an AND node (after [3]). A decision statement can be a predicate or a mathematical expression.

(ii) *Decision tables* provide a popular method of presenting complex engineering data, although they normally require a special preprocessor program with implementation for process planning.

(iii) *Artificial intelligence*: Knowledge involved in process planning will relate to components as declarative knowledge and processes as procedural knowledge. Several methods are available to represent declarative knowledge, such as first-order predicate calculus (FOPC), frames and semantic networks. A wide variety of statements can be expressed in FOPC, by predicate, variable, function and constant symbols. For example, the 'atomic' formula-Depth (Hole (X), Y) represents the depth of hole X as Y units. Depth is a predicate symbol, hole a function symbol, X is a variable symbol, and Y is a constant.

Procedural knowledge can be represented by production rules such as IF (condition) THEN (action) statements, similar to decision trees or tables. In a production rule, a condition can be a conjunction of predicates. For example (after [3]),

$$\left.\begin{array}{l} (=(\text{shape}\&x)\text{hole}) \\ (>(\text{DIA}\&x)0\cdot0) \\ (\le(\text{TP}\&x)0\cdot002) \end{array}\right\} \quad \text{condition of production rule}$$

$$=\Rightarrow \qquad \text{—representing 'THEN'}$$

$$(\text{rapid–travel–out}\&x)$$
$$((:=(\text{TP}\&x)0\cdot01)0\cdot8)$$

in which '$\&x$' is an entity, and actions of the rule are assigned rapid–travel–out to surface $\&x$ and TP (true position) is $0\cdot01$. The constant $0\cdot8$ in the action represents a weighting factor or preference of accepting these actions.

A mechanism is next required to apply the rules to the descriptive knowledge, and provide the process plans. A typical control strategy uses AND/OR solution graphs to proceed via decomposition from a goal to the facts (as backward planning) or from the facts to the goal (as forward planning) using production rules. Fig. 6.9 (after [3]) illustrates an AND/OR graph, in which the goal is to find A using facts D, E, F and G, and rules (1-B and C$\Rightarrow$A), (2-D OR E$\Rightarrow$B) and (3-F AND G$\Rightarrow$C). A significant portion of process planning is not quantifiable, and the AI approach can provide a suitable representation and support the decision making process.

Section 7.2.10 includes further discussion of knowledge based systems.

6.3.2.2.2 Process planning systems [3]. Most currently available process planning systems are of the variant type, including CAPP, MIPLAN, MITURN, MIAPP, ACUDATA/UNIVATION, CINTURN, COMCAPP V etc. The relatively few generative systems include CPPP, AUTAP and APPAS.

CAM-I CAPP (1975) is probably the best known automated process planning system, developed by McDonnell Douglas Automation Co (McAuto) under contract from CAM-I (Computer Aided Manufacturing–International). The CAPP system is based on group technology for variant planning, and is a data base management system in which the coding scheme for part classification is added by the user.

MIPLAN and MULTICAPP (1980) are variant systems, similar to CAPP, and use the MICLASS coding system for part description. The retrieve process

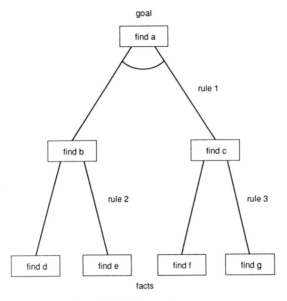

Fig. 6.9   Control strategy using AND/OR graph

plans to be edited by the user are based on part code, part number, family matrix and code range.

APPAS (1979) (Automated Process Planning and Selection) is a generative system for detailed process selection, using decision table logic. Detailed machine surface information, including holes with special features such as oil grooves etc. is described by a special code using a data string of attributes. The system will select feed rate, cutting speed, tool type and cut dimensions for each tool pass. Machining parameters can be optimised and time and cost estimates provided by the process plans.

AUTAP is a comprehensive planning system of German origin, with facilities providing raw material selection against part description, process selection, process sequencing, machine tool selection, tool and fixture selection, and part program generation. The process sequence and machine tool selections are based on decision table logic, with variables supplied by the user.

CPPP (Computerised Production Process Planning) was developed for planning cylindrical parts, and specifies machine tools, cut requirements, reference surfaces, clamping surfaces, and tool and machining parameters using an operation matrix formed from part surface and operation data.

TIPPS (Totally Integrated Process Planning System) is a research-based system that evolved from APPAS and CADCAM (1981), and integrates CAD and generative process planning. The TIPPS structure is illustrated in Fig. 6.10 (after [3]), with the logical divisions of process planning broken into functional modules. TIPPS input is received from a CAD data base, including information such as size, tolerance, and surface finish, linked to corresponding component faces. The surface file and CAD data base links to a process selection module,

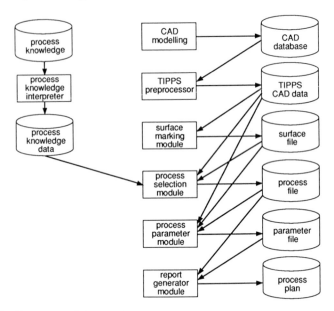

Fig. 6.10   Structure of the totally integrated process planning system (TIPPS)

and process parameters such as feed, speed and machining time are determined by the process parameter module using inputs such as material type, hardness, tool material, tool diameter, surface shape and dimension.

6.3.2.3 Tool requirement planning.   A detailed tool requirement plan incorporating the availability of tools, tool magazine capacity and part grouping related to tool types will be required for effective operation [93]. The processing of a part requires all the necessary tools to be available in the tool magazine, and because of limited capacity this will require tool changing with an increased loading of parts into the system. This raises the problem of selecting jobs to be loaded based on tools required and availability, machine capacity and due dates, for which a linear programming model has been proposed [94, 95]. The overall tool set up problem has also been formulated using a 'tool graph', related to part types and operations, and structured as sub-problems concerned with the choice of tool sets in cribs to minimise transportation system workload, transportation time and tool change time [96].

   With demands for an increasing number of tool types required for an increasing range of parts, tools, fixtures and parts must be scheduled jointly, and it will be necessary to take account of tool changing required owing to tool wear. Results have indicated that tool changes needed due to wear are more significant than changes required owing to product variety, which indicates the importance of incorporating changes due to wear in scheduling rules. It may also be necessary to reduce product variety to simplify the tool management problem.

6.4 Production scheduling

Production scheduling is concerned with implementing the working schedules produced by the planning function, and allocating resources over time to produce a set of part lots. It is by definition a 'push' system which defines a timetable for the immediate future in order to implement a previously defined input sequence of parts at each machine [89].

The production schedule must consider various operational constraints including:

- *Production constraints*: such as machining times affected by availability of parts and raw material, and due data requirements
- *Resource constraints*: limited capacity of operational equipment, such as machines, fixtures, transport system etc.

The shop floor is a dynamic environment with unpredicted changes, such as tool degradation and machine breakdown, occurring frequently. This requires predictive planning to enable previously scheduled activities to be revised in response to changing conditions [97]. Dispatching is also needed to cater for queueing time which is the largest component of manufacturing lead time—the removal of queues removes the need for dispatch rules. Effective scheduling will lead to reduced inventory and storage costs, improved work flow with less work-in-process and better machine utilisation.

6.4.1 Scheduling strategies
The algorithmic approach to scheduling using mathematical programming, develops a feasible solution which maximises (minimises) an objective function subject to a set of constraints. Many techniques have been developed using, for example, linear programming, dynamic programming, integer programming, etc., and search procedures such as branch and bound methods [98]. The techniques are often tested using queueing network theory and simulated using low-order discrete time dynamic models.

The overall problem is combinatorial and difficult to formulate and solve using a meaningful mathematical representation of the production system, incorporating a mix of machines and a relatively large number of parts and tools, and feasible analytical solutions are only possible for small-scale problems [25c, 24a]. An exhaustive enumeration for selecting an optimal sequence of jobs to be processed is also not possible because of the computational problems involved—there are $(J!)^M$ sequences for a problem of J jobs each requiring an operation on each of M machines. The problem can be simplified by generating only a small set of desirable sequences which take account of precedence relations of job sequencing and machine ordering [99].

There will also be many conflicting objectives relating for example to maximum throughput, balanced machine work loads, due dates and minimum tool changing and handling [25b, 100], and the system will be subject to random disturbances, the effects of which can only be counteracted by restricting capacity requirements within the planning and scheduling routines. The representation of uncertainty in demand and equipment failures usually leads to an intractable problem [2]; and much of the theory developed in the

OR field has been found unsuitable for implementation and management support, because of data requirements and computational problems and since much judgment and intuition is required in practice.

6.4.1.1 Job shop scheduling.   The job shop usually contains a large number of jobs each with different routes and requiring operations to be performed in a specified sequence. The problem is to assign priorities to each job queueing at each facility, and to optimise a set of criteria [101]. The problem is complicated since work flow is not unidirectional, and machine operations differ according to jobs to be processed [78].

Batch manufacturing systems require scheduling procedures that can typically: prevent bottlenecks, minimise inventory, control work-in-process levels, minimise manufacturing batch lead times, and maintain acceptable machine utilisation levels, subject to meeting due dates, minimising costs and maximising profit rate.

Many different practical scheduling strategies or local priority rules have been considered and implemented in manufacturing systems. They are relatively easy to formulate but take no account of global status information which could give improved system performance. They include [23, 102, 103]:

- A shortest processing time schedule (SPT)
- Random routing or routing to nearest free machine of correct type
- Due date priority
- Balanced machine work loads
- Use of production ratios
- A periodic input sequence

Scheduling can proceed in either a forward or backward direction. Forward scheduling proceeds from a start date towards a completion or due date, using scheduling rules based on forecasted data and operation time steps. Backward scheduling moves from a due date to an order start date, using scheduling rules to determine times for operations [97].

A mathematical formulation incorporating a heuristic solution of the large scale problem involved in workshop planning and scheduling is detailed in Reference [104], using the interactive programming language APL to handle multi-dimensional arrays. Group technology analysis is used to assign parts to families, and parts within a family are produced according to certain common manufacturing characteristics within specific cells. Operation times are coded in three-dimensional tables relating parts, families and machine tools, and a cell loading function allocates requirements to achieve a uniform loading subject to capacity constraints on tool magazines, manpower etc.

Sequencing of production involves the definition of the order in which operations should be performed, and may need to be considered separately from scheduling which defines the timetable for a pre-defined sequence [89]. This may be necessary, for instance, to adapt a previously defined sequence because of a broken tool replacement, and because of bottlenecks changing from one period to another with a large number of different parts.

In the modern production strategy, optimisation is basically concerned with operations on constraints, determined by dividing the planning tasks into subproblems using say a GRAI decision network analysis method [105] (see

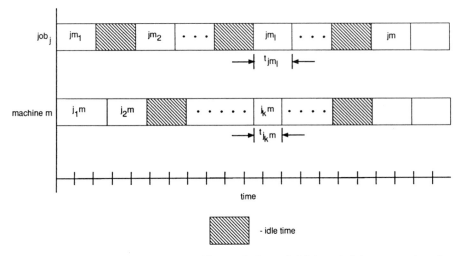

Fig. 6.11 GANTT chart for machine ordering of job *j* and job sequencing for machine *m*

Section 7.2.1). The constraints can be used to direct the search for the schedule to implement, and conflicting constraints can be relaxed as the search progresses to obtain an overall solution—termed 'constraint directed search' [106].

In practice, the scheduling problem will often be solved using heuristics, with little account taken of constraints such as precedence of operations [107]. One such strategy is based on determining the 'bottleneck' workstations and optimising the order sequence/throughput for these stations using realistic process parameters such as set-up times, maintenance times, and carrying orders to the previous planning window if capacity is exceeded [105]. Orders to other workstations are then scheduled starting backward (or forward) from the bottleneck workstation. The heuristics to solve the combinatorial optimisation problem can be based on the 'travelling salesman problem' or other graph search methods to derive the planning sequence. A relatively wide range of knowledge based scheduling systems, implemented mainly as prototypes, is reported in Reference [116].

One of the simplest and most widely used models for scheduling is the Gantt chart which provides a graphical representation of resource allocation with time, as illustrated in Fig. 6.11 (after [108, 99]). It is a horizontal bar graph plotted on a time scale, showing the activities of the job or machine operations involved and their interrelationships. The information may also be represented as Gantt matrices, with elements defining machine and job indexes. The charts record the sequence of steps and time needed to manufacture the end product. They are reasonably efficient for low-dimensional problems, but become unwieldy for medium- to large-scale problems [97]. CPM (critical path method) and PERT (review technique) charts have been developed to extend

Gantt charts for more complex problems. In the CPM method, activity times are analysed using network diagrams to give start dates for each activity and path times.

6.4.1.2 Dispatching rules.   In shop scheduling, the objective is to utilise the production capacity effectively and satisfy customer demands on quality and delivery. The effectiveness of the shop performance may be related to criteria such as minimum cost, minimum in-waiting inventory, minimum completion time of all jobs, maximum profit, maximum utilisation of machines, or ability of meeting due-dates. Reference [109] provides a survey of over 100 priority dispatching rules.

Some typical criteria can be characterised as follows [99]:

Completion time criteria: Job completion time is the most frequently used criteria, and can be represented as

$$C_j = R_j + \sum_{l=1}^{M} t_j m_l + \sum_{l=1}^{M} w_{jm_l} = R_j + T_j + W_j \tag{1}$$

where t_{jm_l} is the processing time of job j on machine m_l

w_{jm_l} is the waiting time of job j before being processed on machine m_l

and R_j, T_j and W_j are the release time, total processing time and total weighting time of job j, respectively. The completion time of a set of jobs J ($j = 1, 2 \dots J$), is the maximum job completion time, or schedule time, stated as

$$C^+ = \max_j [C_j]$$

Times R_j and T_j are known in advance, and thus a sequence is optimal with respect to the schedule time if the total weighting time of all jobs is a minimum. *Utilisation criteria*: The total idle time experienced by all machines is given by

$$I = \sum_{m=1}^{M} \sum_{k=1}^{J} I_{jkm} = \sum_{m=1}^{M} I_m$$

where I_{jkm} is the idle time of machine m before processing job j_k, and I_m is the total idle time on machine m. The idle time on a machine is dependent upon job completion time and the availability of the succeeding job. The sequence will be optimal with respect to the schedule time if the total idle time on all machines is a minimum. On the other hand, consideration must be given to the effects of eliminating machine idle time, which can create large in-waiting inventory.

If all jobs are assumed to be available simultaneously, the shop utilisation can be defined as

$$U = \left\{ \sum_{m=1}^{M} \sum_{j=1}^{J} t_{jm} \right\} \Big/ F^+ = \left[\sum_{j=1}^{J} T_j \right] \Big/ F^+$$

where $F^+ = (C - R)^+$ is the maximum job flow time. The mean shop utilisation is

$$\bar{U} = U/M = [J\bar{T}]/\mathrm{MF}^+$$

where $\bar{T} = T/J$. Thus the mean shop utilisation is inversely proportional to the maximum flow time, and any dispatching rule that minimises the mean $\bar{F}$, in the dynamic case, also maximises $\bar{U}$.

This type of analysis can be used to obtain relationships between the mean flow time $\bar{F}$, the mean number of jobs in the shop $\bar{N}_s$, the mean completion time $\bar{C}$, the mean waiting time $\bar{W}$, and the mean lateness $\bar{L}$, for known values of $\bar{T}$, M, $\bar{R}$ and $\bar{U}$; and any scheduling procedure which minimises $\bar{F}$ also minimises $\bar{W}$, $\bar{C}$, and $\bar{L}$.

Changeover criteria: Machine changeover is often considered to form the dominant criteria for determining a sequence. The total changeover time for all machines is expressed as

$$H = \sum_{m=1}^{M} \left[\sum_{k=1}^{J} t_{j_{k-1}j_k m} + t_{j_J j_0 m} \right]$$

where $t_{j_{k-1}j_k m}$ is the time required to change the machine set up from processing job j_{k-1} to processing the succeeding job j_k, on machine m, and $t_{j_J j_0 m}$ is the time required to return machine m to its initial setup after completing the last job in the sequence. In certain cases it may be more appropriate to consider the number of setups instead of the setup times.

Simple dispatching rules can be based on selecting a job on a first-come, first-served basis, with the shortest processing time or earliest due date, or on the basis of the remaining number of operations or queue lengths. Periodic scheduling involves scheduling a small representative sample of the day's mix, or 'minimal part set' (MPS), and repeating this at regular intervals until the total mix has been produced [110]. The problem is to determine a schedule with a period governed by the bottleneck machine having the largest MPS workload. Dynamic balancing with minimal queueing can be achieved by using a loading sequence which will equalise approximately the cumulative work loads of all machines.

A knowledge based system ISIS has been developed at Carnegie–Mellon University for job shop scheduling, using constraint-directed reasoning and heuristic search [98, 111–114]. ISIS is a production management system which determines the start and completion times for operations and selects machines to perform given operations using information about processing rates, current work load and processing capabilities. DISPATCHER [115] has been implemented in OPS 5 as a knowledge based software system for dispatch of current orders, using a database of products, manufacturing operations required and available, order priorities and due dates. Other knowledge based systems for the scheduling and integrating of automated manufacturing systems are reported in [198–201].

6.4.1.3 Load balancing. Load balancing of the many workpieces to be machined on different machine tools will be required to utilise manufacturing equipment to its full capacity [10]. Many balancing algorithms have been developed based on heuristic models, lexicographic and permutation search, and branch and bound methods [108].

Assembly line balancing can also be attempted to equalise the workload of adding components to a product proceeding down a production line. The work element task times for different components may vary considerably, resulting in different workloads for the assemblies. The line balancing problem is concerned with creating equal-length station times, and is characterised in terms of assembler station times, assembly line cycle time and output, and the number of work elements assigned to a station. Numerous algorithms have been developed using heuristic methods, linear and dynamic programming and network methods. Heuristic methods produce a sub-optimal solution in successive steps and require the least amount of computation time, and are considered to be the most important approach.

6.4.1.4 Robustness.   Robustness of the manufacturing system, or the ability to maintain stability and desired performance, will be affected significantly by the operating logistics and structure of the system, and particularly by the sensitivity or validity of any model-based procedure used to plan, schedule and control the resources involved.

The problem has been studied extensively in the control systems field with particular reference to the robust design of continuous-and discrete-time control systems, to ensure that closed loop stability and performance are insensitive to the unknown components of the system dynamics [117–120]. A well known strategy is self tuning and adaptive control which attempts to reduce the effects of noise and changes in system parameters on system response using simplified models. Adaptive systems can, however, become unstable through the effects of neglected dynamics and unmodelled disturbances in the system design, even if very small.

In the manufacturing system, similar conceptual problems arise, and particularly the increasing extent of automation and integration will make such systems less tolerant of errors. Improved knowledge based designs, using more accurate model-based adaptive techniques incorporating state tables and the monitoring of current system states, will be required. The systems must have the ability to handle and interpret a wider range of sensory operating data to infer the state of the system and its environment, and to activate recovery action using rule-based knowledge incorporating system goals [107].

6.5 Production control

The traditional definition of production control at the higher factory level encompasses functions for planning and control, including:

- Forecasting product demand
- Master scheduling
- Material requirements planning (MRP)
- Order release
- Shop floor control, inventory management

Forecast models are used to project demand over a range of short- and long-term time horizons, to provide inputs to the Master Schedule which determines broad future capacity requirements (see Section 6.3.2.1). The MRP

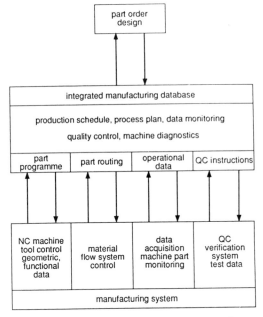

Fig. 6.12 Data flow and control information in an integrated manufacturing system

function determines the required quantity and timing of each product item and when production or ordering of subassemblies should start based on the date of the final assembly. The operation sequencing of work centres is then determined, and the order release function is activated to initiate material purchase and production.

The low level shop floor control function executes the production plan in real time, by initiating and tracking the operations of the machine tools and material handling system, and collecting data concerning product flow, machine performance, tooling, quality control and inventory. The various data flow and control information requirements for integrated production control are outlined in Fig. 6.12 (after [10]).

6.5.1 Cell control
The cell level control system translates long term production requirements into shorter term dispatch commands which are used to co-ordinate and supervise the machining centres and material transport systems. It communicates continuously and in real time with the machine controllers to route parts through a timed sequence of selected machines, reduce unnecessary idle time and to meet production requirements. The cell controller will need to have the capability to reschedule machining operations and re-route material and components following for example machine failures.

The advanced cell computer will incorporate a model of the virtual manufacturing system to simulate the machining operations and part flows, which will enable it to analyse and react to the current status of work on various machines, according to set goals and the actual and planned job times in the contents of a state table [107]. It can also be used to synchronise part flow and to ensure that tools are provided to produce a particular workpiece and that the tool paths are collision free. The cell controller will also supervise and monitor the inspection/quality control systems and the status of tool wear and necessary tool changes.

6.5.2 Machine level control

Machine level control involves real-time control of the production processes in accordance with the scheduling requirements and performance criteria. A multitude of simple control and monitoring functions will be performed by computers and programmable logic units, including:

- Activating and monitoring the control of tool paths, workpiece movements between stations and movements of actuators, relays, valves etc.
- Monitoring completion of work and occurrence of abnormal operation
- Implementing techniques for handling system breakdowns, and developing new schedules
- Tracking location of all workpieces and fixtures
- Tool control and management
- Scheduling and monitoring maintenance requirements for each unit
- Activating and monitoring component inspection policies
- Investigating sources of process errors, e.g. machine or pallet misalignment, tool wear, swarf problems etc.

In particular, NC machine tool handling and control of position, feed and tool speed will be required to position and guide the machine tool to produce a workpiece with a high surface finish and minimum dimensional error. Open and closed loop control configurations can be used for tool positioning; the closed loop control for counteracting the effects of nondeterministic disturbances and sudden overloads by sensing and compensating for deviations between desired and actual tool position.

6.5.2.1 Adaptive control.   Adaptive control of NC machine tools offers, in principle, the possibility of adjusting machining conditions in real time to counteract the effects of unpredictable events such as changes in material properties and cutting forces, premature tool wear or tool and component failures. This requires the sensing of changes in working conditions and the optimisation of some measure of performance such as production rate or material removed per unit of tool cost [28].

Sensors will be required to monitor on-line, or estimate indirectly, parameters and conditions such as tool forces, spindle torque, motor load, tool deflection, machine and tool vibration, machined surface properties, cutting temperatures, tool wear etc. These will be input to the adaptive control system and processed in real time to determine such outputs as spindle speed, feed rate (which has the greatest influence on surface finish) or slide velocity, for optimum machining conditions. The concepts of adaptive or self-optimising control in NC machining are illustrated in Fig. 6.13 (after [28]) and Fig. 6.14 (after [10]).

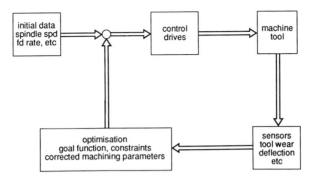

Fig. 6.13   Adaptive machine tool control

Closed-loop adaptive control, although providing optimum conditions with increased tool life, requires complex data processing and a highly sensitive sensory system, and will involve expensive investment. Open-loop adaptive control, with off-line settings derived using standard tool life equations and experimental data to determine machining parameters and optimum-seeking machining conditions, can also be beneficial.

An adaptive control optimisation (ACO) system, able to detect tool wear, tool breakage and other parameters, has been researched for application to an end milling process [121]. Adaptive control constraint (ACC) systems, in which a parameter such as cutter spindle bending or motor torque is constrained to a particular value by self regulation, have been used commercially. Such systems are not true adaptive systems and do not have the ability to react to changes in component geometry under machining conditions, which require sensors able to detect geometric features at machining speeds which is extremely difficult.

The identification task in adaptive control will need to evaluate current performance on-line by direct calculation or by means of a dynamical model. The model will usually incorporate correlation or search techniques for on-line

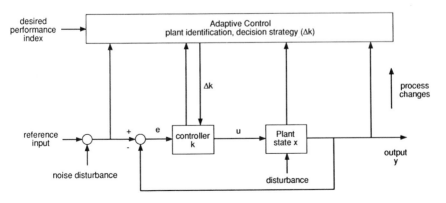

Fig. 6.14   Adaptive control system

parameter estimation or detection of structural changes to reflect process conditions, such as tool wear, or operating constraints, such as minimum or maximum speed, feed, torque and force.

The decision strategy determines the optimal control set point for desired performance quality, using inputs of identified process changes and the desired performance index, as indicated in Fig. 6.14. In the general case, the process will be represented by a set of ordinary differential equations, and optimum settings of the process and controller parameters can be determined in principle by optimisation of an integral form of goal function [28].

Adaptive control can reduce the problems of part programming and program testing required to achieve satisfactory cutting conditions, by transferring the problem of calculating feed and speed of cutter path and of constraint limits for forces, torques and cutter geometry, from the programmer to the controller. The design of an adaptive control system incorporating nonlinear control loops and process identification, using fast and reliable intelligent sensors and online tuning algorithms, possibly with learning and self-optimising capabilities is, however, complex, and the difficulties may prevent reliable implementation.

Chapter 7
Modelling aids

7.1 Introduction

Many model-based solution aids are available to support the design, planning, scheduling and control problems in manufacturing, including the use of discrete event simulation, group technology, computer aided process planning, queueing networks, mathematical programming, perturbation analysis, Petri nets and artificial intelligence.

Mathematical models are useful in identifying key factors affecting system performance and interactions, and, in fact, all activities in the integrated manufacturing system need to be model-based to provide desired reference levels. Models can be used to explore the robustness of the system configuration, and particularly the influence of different material handling systems, location and capacity of inventory banks, required number of pallets, and tool magazine capacities. They can be simple in form and operated on-line as part of the decision support system, or used off-line as detailed representations for investigating the effects of changes in control algorithms.

Models have application in the design problem to determine, for example, the appropriate type and number of machine tools, and the capacity of buffers and the material handling system. In planning they can be used to determine the optimal partitioning of machine tools into groups, and the allocations of pallets and fixtures to part types and the assignment of cutting tools to magazines [96]. They also form the basis of many scheduling algorithms to determine the optimal input sequence of parts [55].

Various graphical techniques have been developed and used as aids for decision making in production control systems. These include:

- PERT networks and extensions in general activity networks (GAN), and GERT models
- GRAI nets (Graphes à Résultats et Activités Interreliés), i.e. graphs with interrelated results and activities
- IDEF: Integrated computer aided manufacturing DEFinition models
- Petri nets
- SADT: Software Analysis and Design Technique models

The basic principles of GRAI nets, IDEF models, Petri nets and SADT models are discussed in Sections 7.2.1, 7.2.2, 7.2.9 and 2.4, respectively.

7.2 Model types

7.2.1 GRAI

A conceptual generic model has been developed for analysing, designing and specifying production management systems within an integrated framework which can be adapted to any kind of manufacturing system [100]. The model incorporates decisional, physical and informational subsystems, and is decomposed into levels each containing a set of decisional centres which control the physical system. The method uses graphic tools, including a GRAI grid similar to that illustrated in Fig. 7.1 (after [100]).

The grid represents an activity–time map, in which a top-down type of analysis [122] proceeds by drawing up a frame, with the vertical levels representing horizons or periods of decision making in a hierarchical ordering of activities from factory to machine levels, and the columns the various functions of production management, such as planning, design, purchasing, resources, control and delivery. Each decision centre with a local controller is linked to others involving different functions and horizon periods, which enables structural defaults such as wrong co-ordination between decision centres to be detected from the grid.

A bottom-up type of analysis can proceed as a second phase by checking each decision centre in detail using the GRAI net, and considering issues such as input–output rules, dynamic aspects of production, variables used and constraints existing in the decision centres.

The GRAI model provides a valid tool for the design of production management systems, using a graphical framework which enables the effects of constraints, subgoal interaction and the lack of co-ordination between decision centres to be assessed.

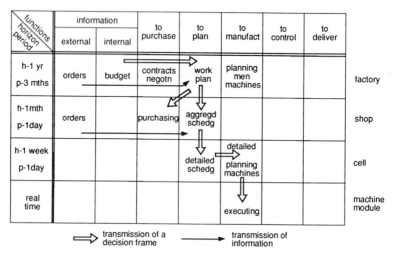

Fig. 7.1   Elements of a GRAI grid

7.2.2 IDEF

IDEF is a well known structured modelling technique for particular application to large complex manufacturing systems, developed within the US Air Force programme for Integrated Computer Aided Manufacture (ICAM) of major air frames [20, 123].

The methodology consists of three basic levels which are used to define the static, informational and dynamic relationships within manufacturing systems [9d]:

IDEF(0)—lowest level—used to define the static functional relationships for the activities in manufacturing systems

IDEF(1)—second level—used to provide an information system model to support the functional model IDEF(0)

IDEF(2)—highest level—a model simulation technique for studying the dynamic behaviour of the IDEF(0) and IDEF(1) models

The building block of all IDEF(0) models interconnecting functions at any specific level in the manufacturing system is illustrated in Fig. 7.2a (after [9d]). Fig. 7.2b illustrates the data flow associated with a particular functional relationship for a typical IDEF(0) model. The IDEF(0) functional model for the total system design is hierarchical in structure, which enables the system to be decomposed into more detailed functions in an ordered and logical manner.

The technique incorporates a blend of structured analysis concepts (see Section 2.4), which allows discrete activities and their associated mechanisms, inputs, outputs and control requirements to be identified, from the upper strategic planning level to the lower operational levels. It enables existing facilities to be investigated and rationalised to improve company performance, and the design of new facilities to be studied to meet specific objectives.

7.2.3 Dynamic production model

Continuous- and discrete-time modelling of the dynamic behaviour of manufacturing systems, incorporating management decisions reacting to customer demand, and the effects of distributors and stock levels, has been investigated [124] using Industrial Dynamics type models to optimise material flow. A typical block diagram representation of the dynamic model is given in Fig. 7.3.

The model includes weekly scheduling based on current sales, stock adjustment to preset levels and rescheduling with demand fall-off, with trial-and-error parameter setting based on previously recorded performance. The model considers dynamic behaviour in terms of production control, manufacturing capacity requirements, inventory holdings and completion of orders, and demonstrates that improved performance with reduced lead time could be achieved by eliminating the lag between scheduling procedures at the factory and in distribution. It demonstrates the existence of order flow amplification and over-reaction to demand in production–distribution chains, and the volatility in stock levels resulting with production ordering based on the feedforward of average sales levels.

An industrial production process has also been modelled as a discrete-time system with states representing rates of flow of parts or subassemblies at various workstations, parts awaiting processing and finished product inventory level [125]. The model includes a disturbance input due to sales, and the control

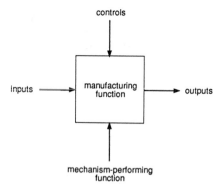

(a) basic function building block

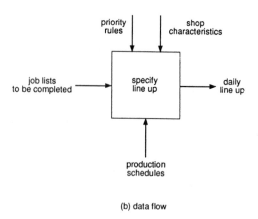

(b) data flow

Fig. 7.2   Building block and data flow for a typical IDEF(0) model

variables are the man-hours scheduled for various work processes. Unconstrained dynamic programming is used to minimise a quadratic performance criterion, giving a standard LQP control solution for maintaining state and control variables close to desired values.

7.2.4 Generalised modelling
An attempt to produce a generalised model representing the dynamics of production systems described by a small number of discrete events, and possibly including a wide range of production systems with different structures and rules for control, has been reported [126].

Each production unit is represented as a service stand of resources (equipment etc.) with input and output buffers, as illustrated in Fig. 7.4a. The lowest (zero) level work centres are linked to form a first level functional or group

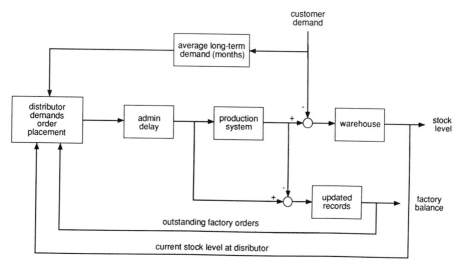

Fig. 7.3 Simplified aggregated dynamic production model

layout, and a set of these units forms a second level production unit, and so on. An example of a typical production structure is illustrated in Fig. 7.4*b*. The resources of the production units may be represented by a few essential discrete event states such as start and end of service, interruption of service or waiting, breakdown, and can incorporate quantity and reliability characteristics.

A general systems approach to the modelling of a production function for sequential decision making in a job shop situation has been considered [127]. The structure of the production problem is defined in terms of a three-dimensional vector space (WHAT, WHERE, WHEN). WHAT represents the parts list and WHERE the list of all resources, and the (WHAT, WHERE) relationship defines the technological function or the fundamental expression of the production organisation. The production function with structural levels is interpreted empirically in terms of a hierarchy embedded in a Taylor series expansion. This forms the basis of the (WHAT, WHERE, WHEN) decision support system for the production function and involves sequential decision making.

A cybernetic modelling of manufacturing processes, with application in CAD systems, is proposed in Reference [53*a*]. The manufacturing process (MP) is represented by a set of characteristics

$$MP = \langle H, F, S, Z \rangle$$

where H is a set of interacting environmental relationships, F is the MP function, S the MP structure, and Z is a set of parameters. The complex is considered as a cybernetic system of control, represented in the form of Fig. 7.5, with material blanks and control information forming the system inputs.

The model $H(Q,U)$, where Q represents nodes and U is a set of arcs linking nodes, is used to describe the control of production complexes and the

(a) Basic module

r^l_i = resource (l=unit level, i=identity no)

S_1, S_2 = buffer storage

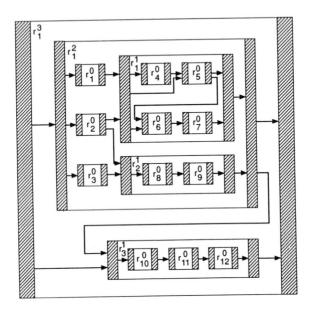

(b) Third level production structure

Fig. 7.4   Modular production structure

interactions between departments or processes. The transformation of the blank to the finished part is specified by a relationship

$$\varphi_{MP}: C_0 \rightarrow C_f$$

where the blank state C_0 is described by parameters characterising the shape of the blank, its material and mechanical properties, and the final state C_f specifies the shape, accuracy and physical properties of the manufactured part. The assembly process

$$\varphi_{AS}: \{DT_i, AS_j\} \rightarrow AP$$

defines the transformation of a set of parts DT_i and assembly units AS_j into an assembled product AP.

The operations in manufacturing, involving hierarchical, functional, spatial and time relationships, are specified by graph structures S_j, S_f, S_p and S_t,

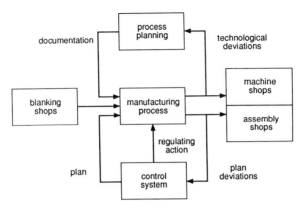

Fig. 7.5 Manufacturing process

respectively. The hierarchical structure has been used to code machining operations and control instructions for NC machine tools. The transformations of intermediate manufacturing states, characterised before and after by shape, accuracy and physical properties, are represented by

$$\varphi_i : C_{i-1} \rightarrow C_i$$

The state of a part after transformation is then described by the relations, $C_{i-1}\varphi_i$, C_i, $C_i\varphi_{i+1}$, C_{i+1} etc., which form the graph of the MP functional structure $S_f(C, A)$, with nodes representing states of the part C_i and arcs the manufacturing operations $[\varphi_1, \varphi_2, \ldots]$. The functional model $S_f(C,A)$ is used as a basis for developing the output language for a CADCAM system.

The temporal structure of the manufacturing process is specified by a set of relations $A_{i-1}A_i$; $A_{i-1}\omega A_i$; $A_{i-1}\tau A_i$, describing the sequential, parallel and parallel–sequential methods (with time shift τ) used for combining the time-dependent operations. A set of links specifying the time dependent sequence of manufacturing steps forms the graph of the operational time structure $S_t(A, \Omega)$, with node set A representing manufacturing operations and arc set Ω the method of combining steps in time. The graph nodes are linked with the time arcs using the arithmetic and logic operations

$$A_i \cdot A_{i+1} \rightarrow t_i + t_{i+1}$$

$$A_i \cdot \omega A_{i+1} \rightarrow \max(t_i, t_{i+1})$$

$$A_i \cdot \tau A_{i+1} \rightarrow \max(t_i, (t_\tau + t_{i+p}))$$

The time for operations of different structures is calculated from the graph $S_t(A, \Omega)$ as the longest route from the initial to the final node. $S_t(A, \Omega)$ can be used for specifying the order of operations and steps on different machines and configurations, involving sequential and simultaneous machining processes and initial timing shifts.

A functional process specification methodology has also been proposed for application as a computer aided design tool for the control of flexible manufacturing systems, requiring online decisions and rapid system reconfigur-aton to cater for product change and system breakdown [9a]. It is based on an analogy between manufacturing and data processing which relates a data file to a set of objects, a data processor to a machine tool, and an object-flow interpreter to the overall manufacturing process. The proposed methodology incorporates an observer concept to reconstruct the system state vector, using automata activated by transition captors or simulators to represent the unmeasurable state variables. The observer output provides input to the command generator for real time control of the discrete event process (see Fig. 7.6 after [9a]).

Fig. 7.7 illustrates a generalised conceptual arrangement of interconnected cells and machining stations with material and information flows, based on a scattering structure [35], which could be developed to form the basis of an aggregated generic model of an integrated manufacturing system.

7.2.5 Simulation

Discrete-event simulation is useful, particularly at an advanced stage of the production design process, for investigating the behaviour of various feasible scheduling and operating policies and layout configurations. It can be inte-grated with an operational data base, thus permitting a form of material requirements planning to be investigated prior to implementation [10, 128, 25a].

General purpose simulation languages, such as SLAMII, GASPIV, TESS, AUTOMOD/AUTOGRAM, GPSS/H, SIMSCRIPT, SIMAN, SIMULA, MAP/1, MICRONET, GCMS, GFMS, SPEED, Q–GERT etc. are available to investigate the behaviour of discrete event manufacturing systems. A review of simulation packages is given in [18d].

Simulation makes it possible to predict the effects of events occurring during the planning/scheduling horizons, including changing the capacities and speeds of material handling systems such as pallet transport devices and robots, and allows bottlenecks and machine under-utilisation to be identified [14]. It can be used to evaluate alternative maintenance schemes, control functions and decision rules such as first-in first-out or priority to shortest operation time

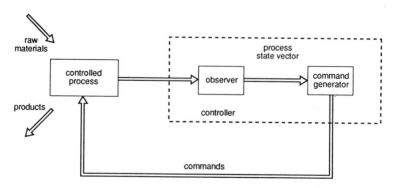

Fig. 7.6   Observer based system control

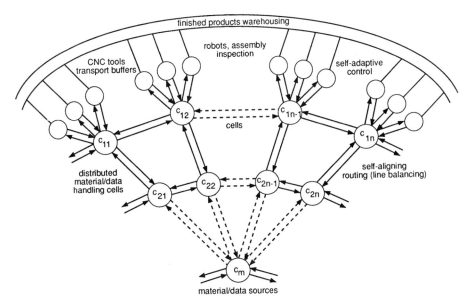

Fig. 7.7 Generic material/component/data flows

[129], using trial-and-error settings. Queue build-ups in the buffer stores can be identified, which will assist in determining capacities required for the various equipment modules [52]. Strong correlation exists between the number of material handling carts and speeds, and between the number of pallets and their effects on workstation buffers, and simulation can be used to obtain an optimal solution by investigating different strategies. Tool transport facilities, tool storage and replacement tool control algorithms can also be simulated [18*d*]. A general methodology for setting batch sizes with a relatively small number of categories and adjusting priority rules using a search procedure to achieve desired performance, such as lowest production costs or lead times, in an event-based simulation model for a large scale manufacturing system, is illustrated in the flow diagram of Fig. 7.8 (after [130]).

Simulation can provide solutions to problems intractable by mathematical analysis, and is the only method of proving that a system design will perform as expected. It does, however, require expertise in output analysis, the design of experiments, and in determining appropriate confidence limits, for model systems incorporating random disturbances such as machine breakdowns, tool breakages and in material supply. Models also are rarely well documented with underlying assumptions, and validation of results is difficult. Simulation model code also is not transparent, other than to the programmer, and simulations can be expensive in their use of computer resources.

Important developments in discrete-event simulation include the enhancement to code generators for constructing simulation programs [18*d*], and exploiting the potential for generating systems software for CIM [128].

7.2.6 Queueing networks

Queueing networks have application for modelling interacting flows of parts competing for machines, at an aggregate level of detail. Average levels of variables such as machine tool processing times and frequency of part visits to a machine are used as inputs to the networks, and outputs include, for example, average values of steady state production rate, mean queue lengths and machine utilisations [55, 131].

The models provide information about average system behaviour over long time periods, and are particularly useful in preliminary qualitative design analysis and in determining the number of machine tool types, the capacity of material handling systems, and in locating bottleneck processes, although they cannot be used for detailed planning. They provide insight into how components interact as parts compete for limited resources in closed queueing network models. They are not, however, particularly useful in studying complex FMS scheduling and control problems, involving multiple parts, and the effects of breakdowns, a limited number of pallets, finite queues at workstations, etc. Modelling has to be on an aggregate level to be analytically tractable, and

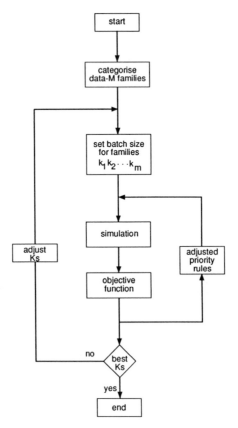

Fig. 7.8   General methodology for setting batch sizes

restrictive assumptions, such as exponentially distributed processing times and infinite queues, are often necessary. Queueing analysis, however, can provide a broad indication of system performance, and can give reasonably good estimates for preliminary decisions without requiring excessive input data and computation time.

The basic theory of closed loop queueing systems was developed by Jackson [25a] and later extended by Gordon and Newell [132] and Buzen [133]. Analytical techniques based on queueing networks have been developed to study the effects of routing policies on the throughput and in-process inventory of a flexible manufacturing system [51, 134—138]. The system manufacturing one or more workpieces can be modelled as a set of stations with servers and queues, and given routings and processing time distributions.

Most queueing models assume a first come–first served (FCFS) queue discipline, although models with state dependent rules for dispatching and routing of jobs allowing operational features have been developed [139, 140].

The most commonly used stochastic models are based on the CAN–Q system (Computer Analysis of Networks of Queues) [18d, 141] in which the process is typically represented as a set of workstations with a material handling device, and workpieces circulate in the closed system with prescribed routing probabilities. The method allows configurations to be evaluated using measures related to the control of part mix and dynamic loading. However, care must be taken in the interpretation of results because of averaging effects and the neglect of interactions.

7.2.7 Perturbation analysis

A discrete event dynamic system (DEDS), such as a manufacturing system with interconnected servers and queues, evolves as a series of discrete events, and the number of states in a typical DEDS can be combinatorially large. The arrival or departure of parts at machines initiates or terminates specific activities, generally randomly, and the discrete event nature of the system requires a representation in an event domain using a counter $k = 0, 1, \ldots$. The state equations of such a system can then be represented [151]:

$$q(k+1) = q(k) + \sigma[q(k), e(k), \omega(k; \theta)]$$

$$e(k+1) = e(k) + \tau[q(k), e(k), \omega(k; \theta)]$$

where $[q(k), e(k)]^T$ is the state vector when the kth event occurs, with components $q_i(k)$ and $e_i(k)$ representing integer queue lengths and times of events occurring as a real-valued variable, respectively. The vector $\omega(k; \theta)$ is a stochastic input parameterised by θ, and $\sigma[\cdot]$, $\tau[\cdot]$ are transition functions.

A linear system theoretic formulation for a restricted class of DEDS has been developed for manufacturing systems without breakdowns, using a minimax algebra which can incorporate control concepts such as transfer functions, controllability and observability [142, 143].

Information concerning changes in the performance $J(\theta_i)$ of a manufacturing system with respect to the decision parameters (θ_i) is generally unknown, and simulation of the complex dynamics involved is usually required to evaluate the sensitivities $\Delta J / \Delta \theta_i$, $i = 1 \ldots N$. These will usually be required both at the design stage to select particular components and also for short term scheduling purposes. The simulation approach, however, is costly, particularly for a number of stochastic realisations of the system, and is infeasible as an online decision tool.

Perturbation analysis (PA) avoids the limitations of both queueing theory and simulation, and can be used to estimate the perturbed performance measures or sensitivities of the system (144—150, 2, 25a]. The method is based on the observation of a single experiment or nominal sample path obtained by simulation or the use of online data, and enables the effects of changing parameters to be quantified without subjecting them to change. This provides guidance for the selection of certain parameter values influencing performance, including, for example, choice of machines and operating times, buffer sizes, numbers of pallets/fixtures etc., and also for online optimisation.

Perturbation analysis computes in real time the gradient or sensitivity vector $\partial J/\partial\theta = [\partial J/\partial\theta_1, \ldots, \partial J/\partial\theta_M]$ by linearisation of the state equations derived for the particular production line. The system performance measure J is in general related to the timing of events such as throughput and weighting time. A nominal sample path for the real or simulated DEDS is observed and specified by a tableau of event sequences ξ_i, $i = 1, \ldots, I$, which summarises the operating history of the system in terms of service times and event durations. A known perturbation is then introduced in the system tableau (e.g. in service time, queue capacity), to give a perturbed tableau of event sequences, which specifies a perturbed path related to nominal conditions using gain update equations, which can be used for predicting behaviour of the DEDS along other sample paths.

7.2.8 Analytical methods

A range of nonlinear integer programming [152], linear programming [153] and dynamic programming [134, 153] techniques have been formulated and used to solve some design, planning and scheduling problems, relating particularly to loading and production rate control at various hierarchical levels. Heuristic algorithms have also been developed for the FMS loading problem [154].

Mathematical programming concerns the optimal allocation of limited resources, such as machine time, tools, pallets, fixtures, to competing activities, subject to a set of constraints such as due dates, routings, part mix etc. [25a]. Constrained optimisation is usually formulated in terms of maximising or minimising an objective function subject to system and state constraints as follows [78]:

$$\text{Maximise: } g(x, y)$$

$$\text{subject to: } f(x, y) = 0$$

$$\underline{x} \le x \le \bar{x}$$

where $\cdot$ $x = (x_1, x_2, \ldots, x_n)$ are controllable decision variables, and $y = (y_1, y_2, \ldots, y_m)$ are uncontrollable parameters, and $\underline{x}$, $\bar{x}$ are lower and upper bounds on x, respectively. The optimal solution or decision rule is given by

$$x^* = \psi(y)$$

Optimisation techniques include:

- Extremum methods—including calculus of variations, Lagrange multipliers etc.
- Mathematical programming—linear, nonlinear, integer, and dynamic programming, etc.
- Network theory—project scheduling techniques such as PERT (Program Evaluation and Review Technique), CPM (Critical Path Method) and GERT (Graphical Evaluation and Review Technique) etc., for shortest-path and maximal-flow problems and line balancing
- Implicit enumeration—iterative procedures for reducing computational steps, including branch-and-bound, lexicographic search, permutation search etc.

The optimal assignment of limited resources to a set of competing jobs, and determination of an optimal mix of products, can in principle be obtained using these various techniques, although there are few reported applications particularly for the FMS selection problem involving multi objectives. The following discussion summarises various methods of mathematical analysis which have been investigated and are worthy of consideration for possible application in production planning and control:

(i) A simplified, decentralised, open-loop control strategy for production planning, with updating of the production flow pattern based on orders and the current state of the system, affected, for instance, by system disturbances such as breakdowns, has been formulated [90]. The planning level provides a reference input for lower-level feedback control, which aims to reduce the effects of disturbances and failures with local replanning over short-term horizons.

The objective of production planning is defined in terms of minimising work-in-process, updating the process plan, and completion of the loaded orders. The model includes representation of failure events, event occurring time, estimated restoring time for an event-affected work stage, order events defined in terms of order release time and due date, and order dimension for a required part type.

Various decomposition techniques are considered for the multi-stage, multi-product problem, including Interaction/Prediction (MRP II push type) using approximate subsystem solutions independently satisfying due dates for each shop; Interaction/Balance (push–pull type) with local plans minimising a weighted cost function and communicating completion and release times to downstream and upstream shops, respectively; Decomposition–Co-ordination (OPT type) and Backward-Conditioned Local-Feedback Decomposition (JIT type).

The push-type procedures, using iteration, provide 'apparent' co-ordination which produces difficulties with large demand variations and require strong co-ordination actions while maintaining small interstage inventories. The pull-type methods have difficulty in handling demand peaks without production forecasts. The mixed 'push–pull' methods, on the other hand, provide a mechanism for integrating local optimisations and have the ability to reduce local inventories.

The mathematical theory of hierarchical multilevel control is developed and discussed in detail in many references, including [155, 156].

(ii) The multi-product, production planning and control problem has also been formulated in terms of separate optimisation activities, in which planning is concerned with scheduling a set of lots governed by required quantity, due date, raw material release time and a given time horizon, and the control problem with machine sharing, lot splitting and job sequencing [157]. Both product- and shop-level-wise decompositions are applied to decompose the problem incorporating time and capacity constraints into a sequence of simpler subsystems.

A projection technique is applied as a solution strategy for the planning-type decomposition problem, and interpreted in terms of MRP II concepts. The products are scheduled independently and the interacting concurrent production capacity constraints are accounted for interactively. A dual problem formulation is developed to induce partial level-wise separation, with each workshop controlled independently and taking account of capacity constraints. Material release times and due dates, as interaction constraints among shops in the same production tree, are satisfied using penalty terms as control variables in the local performance indices.

Concepts of the OPT planning technique can be incorporated by taking account of an expected system bottleneck, and propagating the resulting schedule upstream and downstream along the production stages without violating capacity constraints. Simplified practical implementations are expected to achieve good scheduling efficiency, with reasonable computational effort.

(iii) The joint flow and service control problem in flexible manufacturing systems, involving the simultaneous processing of a relatively large number of part types (of different products or part families) and the sequencing of workstation operations, has been investigated, with the combined problem reduced to a nested sequence of decoupled optimisation problems [36]. These involve:

- flow assignment for each part type to maximise section throughput, based on known service rates, and
- service rate planning to minimise section clearance times, based on a known flow pattern

The system model is formulated using a dynamic balance equation based on workstation service rates, input/output buffer levels, part flow continuity between consecutive stages, and feasibility sets for service rates of both stations and loading devices. The flow-and-service control problem is concerned with determining a constant optimal flow pattern of stage service rates minimising the section clearance time, and a constant optimal flow pattern maximising section throughput, conditioned on a constant pattern of stage service rates. The use of square wave input work demands allows an independent solution of the flow-and-service control problem, with different part type mixing. A nested sequence of performance equations permits a decoupling, with the service control problem solved as a linear minimum-time control problem and the flow assignment as a linear programming problem, with constant input.

(iv) Operational planning functions include: batching, routing, dispatching and sequencing. Considering the first two functions: batching defines the list of parts to be processed in the next time period, and routing assigns a process plan for each part type minimising the production time. The decision processes involved will need to be sequential and iterative because of the problems of global optimisation, and solutions will usually be formulated using linear programming models.

In the FMS batching problem, the objective function can be related to the number of part types assigned to particular routes, weighted according to available pallets, the number of routes, processing times and the number of new tools required. An algorithm has been developed to determine the minimum number of part types required to be processed in the next time period, in order to guarantee due date feasibility, using a linear programming model with constraints on available machine times and tool magazine capacity [89].

In the routing problem, the objective will be to assign a process plan to each part type, and minimise the time required to manufacture one batch, subject to the above constraints and taking into account alternative routes. An iterative optimal solution can be obtained using a linear programming model which not only minimises completion time, i.e. time required to process the batch at the bottleneck station, but also balances the work between the non-bottleneck stations, which improves part flow through the system with smaller queues in front of machines [89]. In sequencing, the problem also reduces mainly to controlling bottlenecks. Other results show that optimal routing is sensitive to the relative work rates of the stations [134].

(v) Part selection requiring high tool variety can have a significant effect on planning and control in FMSs. The simultaneous processing of all part types is constrained by the tool magazine capacity, and sequential processing of selected order subsets using a real-time algorithm has been proposed [158].

The part selection procedure is invoked when:

- The work load assigned to the machine group with the highest utilisation is less than a predicted threshold value
- A large number of empty pockets is available at the workstations
- Urgent orders need to be processed.

The model includes constraints on tool magazine capacity and due dates, but not on machine capacity.

The aim is to minimise an objective function including weighted normalised deviations from mean work loads, mean tool pocket requirements, specified due dates and tool magazine capacity. The formulation introduces a high computational load, and a heuristic procedure using sequential part selection which permits a proper balance between the elements of the objective requirements is proposed.

(vi) Scheduling an FMS with the balancing of several machines processing many parts is a difficult problem, which is further complicated by the disruptive effects of random disturbances such as machine failures and material shortages. A short-term hierarchical-based schedule, using a simple heuristic procedure has been proposed to reduce the effects of major disturbances, using the arrangement outlined in Fig. 7.9 (after [159]).

Production requirements are specified from a higher level, and the schedule, incorporating capacity constraints, builds up inventory to reduce the effects of machine failures. It solves a linear programming problem only at machine

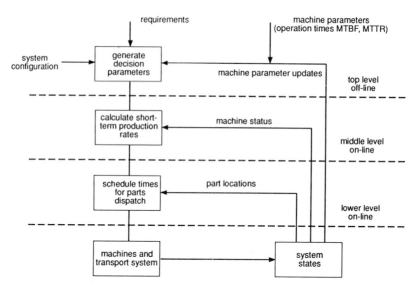

Fig. 7.9   Short-term production hierarchy

failure and repair times, and the scheduling algorithm determines dispatch times for parts to machines to satisfy weekly production requirements. The problem is decomposed into two parts, using (i) high level continuous dynamic programming to determine the instantaneous production rates, and (ii) a lower level algorithm to determine dispatch times.

The policy objective is to determine the production rate vector $u(t)$ to meet production targets, while minimising a total cost function $g(x)$ representing both inventory and backlog costs, subject to machine capabilities. The dynamic programming problem is stated as

$$\min_{u(\cdot)} E\left\{\int g(x(t))\,dt\big|x(0), \alpha(0)\right\} = J[x(0), \alpha(0)] \tag{7.1}$$

subject to a machine constraint set $u(t) \in \Omega[\alpha(t)]$, initial conditions $x(0)$, $\alpha(0)$, and a production surplus vector differential equation defined by

$$dx/dt = u(t) - d(t)$$

where $d(t)$ is a production requirement or demand rate vector, and $u(t)$ a production rate vector. The scheduling policy given by the solution of eqn. (7.1) can be decomposed into the three levels indicated in Fig. 7.9.

Function J is evaluated off-line at the top level, using a quadratic cost-to-go function related to the build-up of inventory. The middle level includes the on-line computation of instantaneous production rates, using the results of the top level computation and a solution given by

$$\min \frac{\partial J(x, \alpha)}{\partial x} u, u \in \mathbf{v}(\alpha)$$

which provides a feedback law when x and α are determined. The production rates u are then translated into actual part dispatch times at the lower level.

7.2.8.1 Entropy in the manufacturing system   The concept of entropy, and its relevance to decision making in industrial organisations involving discontinuous flow processes [160], could have useful applications in the planning and control of manufacturing systems. Such processes with complex interacting subsystems, producing a wide range of products and requiring multi-level decision making, provide conditions for the existence of high levels of entropy within the system.

The industrial system can be considered, from a mathematical point of view, as a heterogeneous structure:

$$\zeta = [\{X, Y\}\varphi]$$

where $X = [x_1, x_2, \ldots, x_m]$ represents the aggregated input signals, $Y = [y_1, y_2, \ldots, y_n]$ the generated output signals, and φ is a set of transfer operators.

Generally, the better organised a system, the lower the entropy level for the entire system, which implies knowledge of information contained within the system. The entropy of the system ζ can be defined by the functional

$$H(\zeta) = \sum_k p_k \log 1/p_k$$

where p_k is the relative probability of malfunction of the component elements of the system. For an industrial system consisting of independent units ζ_i (workshops, departments etc.) and producing a relatively large number of products in many workstations, the total entropy will be

$$H(\zeta) = \sum_i H(\zeta_i)$$

The main problem in the large scale manufacturing system is to co-ordinate and implement efficient solutions in order to reduce the system entropy. This can be achieved by the reduction of entropy in the workplaces, such as with the introduction of NC machine tools (to reduce manual intervention) and automated material handling systems, and generally by the integration of automated equipment. Entropy can, however, be increased through the increased complexity of equipment, and by bad design, increased product variety, inefficient planning, scheduling and control procedures increasing manufacturing costs, random disturbances, excessive inventory and work-in-process, tool wear, poor quality, late deliveries, customer dissatisfaction, and shortage of skilled staff.

Improved system operation with reduced entropy requires calculation of entropy specific to each controlled process and activity, using prescriptive coefficients to classify the elements in terms of achievable entropy levels. Minimum global 'error' in the system operating with a high degree of stability and reliability will be consistent with low entropy. This may require a high level of integrated automation and the development and implementation of complex decision making processes and self-adaptive control systems.

7.2.9 Petri net modelling [161—164]

Petri nets (Pns) provide an abstract model for complex discrete event processes, and although characterised by only a few simple rules they can be used to model a wide variety of discrete event systems and activities including, for example, computer hardware/software systems, communications protocols, distributed data bases, and manufacturing systems. They can be considered as directed bipartite graphs, with directed arcs linking 'places' and 'transitions'. The transition represents an evolution or change of state taking place, and the place represents a condition or state of the system containing tokens associated with resources.

Pns provide a simple and natural representation for a discrete event system, although there is no widely accepted methodology for building Pn models and they can be difficult to analyse. Efficient algebraic techniques have, however, been developed to analyse certain subclasses of decision-free timed Petri nets, which have been shown to be equivalent to linear state equations in a (max, +)-based algebra [165]. These can be used to investigate real-time steady-state and transient scheduling and control problems in manufacturing systems, and can provide information concerning cycle time and production rate, effects of bottlenecks, machine utilisation etc.

Basic definitions and properties of Petri nets are discussed in Appendix 1.

7.2.9.1 Manufacturing system applications.   Petri nets are well suited for modelling the complex interactions between the components of a manufacturing system, and particularly those features involving synchronisation and concurrency of different activities such as control procedures at the work-cell level, which affect significantly overall system performance [163].

They can be used in plant design, say for determining optimal buffer sizes, pallet distribution and material handling capacity, and in assessing the effects of scheduling and sequencing rules on critical resources. Pn modelling and analysis can also provide insight into the real-time behaviour of FMSs over finite time horizons [165].

Various examples of Pns used in manufacturing systems are discussed briefly to illustrate current trends and the potential for further developments and applications.

(i) *A simple machine shop modelling problem [161]*
The actions involved in processing orders for machining and delivery can be defined in a Pn structure:

$$I(t_1) = \{ \cdot \} \qquad O(t_1) = \{p_1\}$$
$$I(t_2) = \{p_1, p_3\} \qquad O(t_2) = \{p_2\}$$
$$I(t_3) = \{p_2\} \qquad O(t_3) = \{p_3, p_4\}$$
$$I(t_4) = \{p_4\} \qquad O(t_4)\{ \cdot \}$$

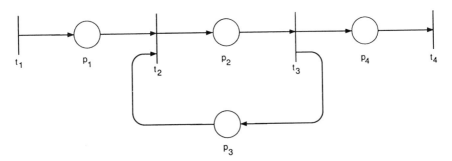

Fig. 7.10 Petri net model for a simple machine shop

which represents the pre- and post-conditions on the corresponding events. The transitions (events) and places (conditions) are defined as:

t_1: order arrives
t_2: machine processing starts
t_3: machine processing complete
t_4: order sent for delivery
p_1: machine shop waiting for work
p_2: order is waiting
p_3: order being processed
p_4: order complete and sent for delivery

The corresponding Pn graph model is illustrated in Fig. 7.10.

(ii) *Planning and control cycle timing [166]*
A Pn representation has been developed to model the relative timing of a number of different levels of planning and control, as illustrated in Fig. 7.11. A five day activity cycle is incorporated within a four week planning cycle, with monthly figures used as targets or constraints for subsequent planning.

(iii) *Three-machine FMS [165]*
A three-machine FMS, processing three different part types with specified part routing, is illustrated in Fig. 7.12. The Pn model is illustrated in Fig. 7.13, with machine tool processing times for each part type operation indicated on the corresponding transitions, and the token types (machine tools and pallets) labelled alphabetically.
It is assumed that:

● Machine set up and transportation times can be neglected
● The product mix is balanced, obtained using a periodic input of parts with the sequence 1–2–3.

The Pn is decision-free, safe, and live for any feasible initial marking. Machine sequencing must be known, and further developments are required to incorporate optimising control rules.

(iv) *FMS information system [167]*
A Pn representation of an information system for an FMS is illustrated in Fig. 7.14. The transitions correspond to major components, with t_1, t_2, t_3

and t_4 relating to the information cells for material handling, manufacturing, inspection and supervisory control, respectively, When t_1 is activated and fired, information concerning the various parts and assemblies used in manufacturing is used to move the components between the various cells and storage areas.

(v) *Inter-operational buffer [168]*
Resource limitations in a production system, for example of working cells, pallets, conveyors, buffer positions etc., and their effect on system performance, can readily be taken into account in a Pn model. For example, a Pn representation of an inter-operational buffer of limited size is shown in Fig. 7.15.

The busy and free positions are modelled by places p_2 and p_3, respectively, and the sum of tokens in p_2 and p_3 represents the number of constant buffer positions. The Pn indicates the particular case of a buffer with three positions, two of which are free (with two tokens in p_3). Transition t_2 models the filling of one buffer position, and requires at least one free position (one token in p_3) to fire.

(vi) *Cell conflict [168]*
The existence of conflict in a production system, in which two cells (C_1, C_2) feed into a single charging machine or resource CS, is illustrated in the block diagram and Pn model of Fig. 7.16a and b. The working state of the cells is

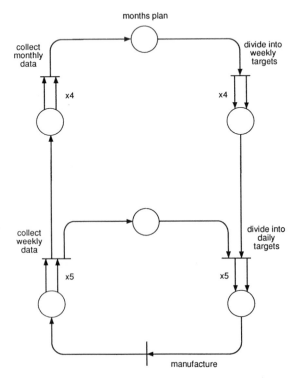

Fig. 7.11   Representation of hierarchical planning and control cycles

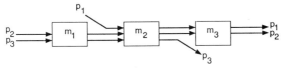

Fig. 7.12 A three-machine FMS

represented by places p_1 and p_5, and the request for access to CS and the busy condition by places p_2, p_6 and p_3, p_7, respectively. The use of resource CS is determined by place p_4 which prevents simultaneous marking of p_3 and p_7, and provides an enabling–disabling facility for the conflicting cells.

(vii) Three cell-buffer complex [168]
A three cell-two buffer production system, with two part types A and B processed in cells C_1 and C_2, respectively, and both in a shared cell C_3 via inter-operational buffers H_1 and H_2, is illustrated in the block diagram and Pn model of Fig. 7.17 *a* and *b*. The transitions (events) and places (conditions) are defined as:

t_1: part A—completion of work on C_1
t_2: part A—filling buffer H_1
t_3: part A—transfer from H_1 to C_3
t_4: part A—completion
t_5—t_8: similarly for part B on C_1, H_1 and C_3

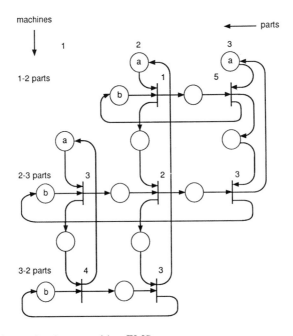

Fig. 7.13 Petri net of a three-machine FMS

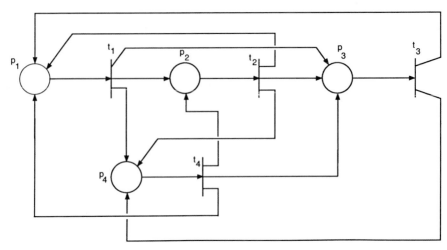

Fig. 7.14   Petri net for FMS information system

p_1: C_1 working on part A
p_2: C_1 waiting to deposit part A in H_1
p_3, p_4: C_1 busy, free positions of buffer H_1
p_5: C_3 working on part A
p_6—p_{10}: similarly for cell C_2, part B and buffer H_2
p_{11}: C_3 idle

In the left hand cycle of Fig. 7.17*b*, cell C_1 can be in condition p_1 or p_2, and may be idle in p_2 if t_2 is not enabled (i.e. only if there are no free buffer positions—no tokens in p_4). The sum of tokens in p_3 and p_4 represent the number of buffer positions. The shared resource C_3 can be either in place p_5, place p_{10} (working part B) or place p_{11}. The conflict situation is controlled by the tokens in place p_{11}, as in example (vi).

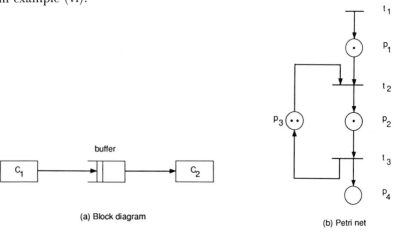

(a) Block diagram

(b) Petri net

Fig. 7.15   Inter-operational buffer

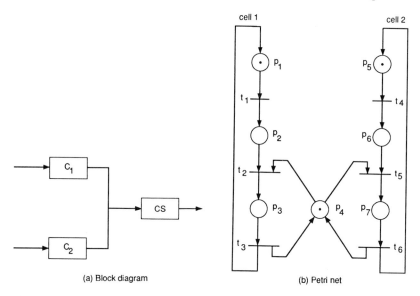

Fig. 7.16 Example of cell conflict with a single charging machine CS

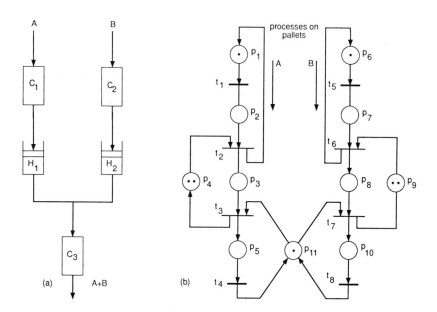

Fig. 7.17 Three-cell/two-buffer production system

7.2.9.2 Coloured Petri nets.   The above examples illustrate the main disadvantage which will arise in the Pn modelling of actual production systems, namely, their size and complexity if the number of machines or parts exceeds three or four. A change in operational procedures such as in the scheduling policy or route definition will also introduce substantial changes in the net structure.

Some of these problems can be reduced by the construction of coloured Petri nets (CPNs), which provide a more concise model obtained by grouping elements with similar behaviour such as machine or part sets [25e]. Analogous elements are distinguished using an attribute or sign called *colour*, which is represented by an identifier. Thus, in a simple two-machine, two-part flowshop, a subset of colours associated with each machine and with each part can be considered with

$$MCH = \{mch1, mch2\}, PARTS = \{part\ 1, part\ 2\}.$$

Compound colours can also characterise the state of a part $\langle part\ i \rangle$ in relation to the machine $\langle mch\ j \rangle$ with the colours

$$\langle part\ i, mch\ j \rangle \in PARTS \times MCH.$$

In a CPN, a place can be marked by a sum of colours, and a subset of colours is associated with each transition. The state of the system then changes on firing a transition with respect to a given colour. The enabling of a transition and the firing rules are governed by linear functions labelling the arcs of the net, and indicate the colours marking each place and the colours which must be added to or removed from a place on firing.

The form of a CPN is shown in Fig. 7.18 for the simple flowshop manufacturing system depicted in Fig. 7.19, producing two parts with two machines MCH 1, MCH 2 [25e]. The CPN includes three places, with markings associated with the set of colours $\langle part\ i, mch\ i \rangle$, in contrast to the 14 places and 8 transitions forming the standard Pn. The scheduling of parts through each machine and routing of each part are defined by the simple functions

$$sch\langle part_i, mch_j \rangle = \langle part_{i+1}, mch_j \rangle$$

$$rout\langle part_i, mch_j \rangle = \langle part_i, mch_{j+1} \rangle$$

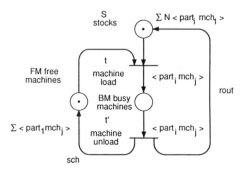

Fig. 7.18   Coloured model for flowshop of Fig. 7.19

Fig. 7.19   Flowshop manufacturing system

The initial marking of place S (stocks) indicates that all parts are waiting in the entry stock for MCH 1, and the initial marking of place FM (free machines) implies that all the machines are free and waiting for a PART 1. Then:

- initial marking enables t with respect to colour $\langle part_1, mch_1 \rangle$
- on firing, PART 1 is loaded into machine MCH 1
- colour $\langle part_1, mch_1 \rangle$ transfers to place BM (busy machines)

The procedure continues with:

- This machine unloads by the firing of t' with respect to colour $\langle part_1, mch_1 \rangle$
- colour $sch\langle part_1, mch_1 \rangle = \langle part_2, mch_1 \rangle$ is added to place FM—machine MCH 1 is ready to receive PART 2
- colour $rout\langle part_1, mch_1 \rangle = \langle part_1, mch_2 \rangle$ is added to place S—PART 1 is ready to enter machine MCH 2

The flowshop policy is thus correctly implemented through the cyclic behaviour of the CPN produced by the successive firings of the transitions t and t'.

Most of the properties of standard Pns, such as boundedness, liveness, mutual exclusion etc. can be generalised for CPNs although further developments are required to extend available analysis techniques.

7.2.10 Knowledge based systems

Manufacturing systems knowledge is relatively unstructured, and decision making involved in design, planning and control is often based essentially on heuristics and the experience and skills of a comparatively few people using subjective judgments and past experience. It is important to capture the skills that are disappearing rapidly, and also to extend the expertise and knowledge relating to new materials and particularly of their properties under extreme operating conditions. The introduction of automation has also stimulated the need to formulate and codify the rules and decision making involved in all the various manufacturing activities [55, 24*b*].

Knowledge based systems use predefined sets of strategies and reasoning methods to solve a problem for which the structure has been defined. An 'expert system' is a special type of knowledge based system which attempts to codify and store knowledge using fuzzy rules to represent the logic used, say, by the experienced planner, and to make it more readily accessible. They have an important role in manufacturing systems, particularly for the many problems involving a complex interaction of logic and conflicting objectives, which are not amenable to mathematical analysis. The expert systems approach is appropriate in process planning particularly, for which there is no systematic methodology to provide a set of algorithmic procedures for process selection and the sequencing of operations, and existing manual methods are subjective and ill codified.

7.2.10.1 Expert systems.   An expert system usually consists of three basic components [122]:

- *Knowledge base*: containing the rule base and the data base of facts describing the state of the domain
- *Inference engine*: which governs the way in which the system searches what it knows, or interprets the rules
- *User–machine interface*: controlling the interactions between the expert system and the user

The types of manufacturing knowledge consist of:

- *Objects*: such as part description, machine capability, queue contents
- *Events*: representing activities such as: part *a* is processed on machine *x*, machine *y* is drilling part *b*

Representation frameworks for problem solving include:

- *State space search*: using stored system states and transition operators
- *Procedural representation*: using stored procedures for specific actions
- *Semantic networks*: knowledge represented in a network of links defining inter-relationships between nodes representing objects, concepts or events
- *Rule based systems* (production systems): knowledge represented as a set of IF–THEN rules
- *Logic based systems* (usually first order predicate logic): knowledge represented as a set of formal logic predicates
- *Special purpose*: such as frames and scripts

In the rule based system, the rules or productions operate with the simple conditional statements—IF this condition holds, THEN this action takes place. Implementation is effected by scanning lists of the IF parts of rules for matching with the THEN parts. The approach has important applications for implementing heuristics representing experience and intuition relevant to decision making in manufacturing [97].

The interpreter decides which production to 'fire' next, and controls procedures using strategies such as:

- Forward chaining along search paths: a bottom-up strategy which establishes a path between an initial state and one or more goal states
- Backward chaining (goal driven) top-down method of inference: solution attempts to reach initial conditions starting with known goal, by deductive examination of selected rules
- Mixed strategies: searching for a path connecting goal with initial state, from both ends of problem
- Problem reduction: using an AND/OR tree and search

The general structure of an expert system is illustrated in Fig. 7.20 (after [1]).

7.2.10.1.1 Applications.   The main developments in the applications of knowledge based techniques have been in certain specific design and manufacturing areas, although little appears to have been achieved in the area of automated process planning linking design and manufacture. This is due largely to the scale of the task and the difficulties of representing and manipulating the knowledge base [74*f*]. Integration of the processes involved requires:

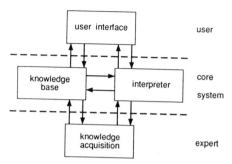

Fig. 7.20   General structure of an expert system

- A comprehensive product model description incorporating a geometric representation appropriate to process planning and downstream manufacturing activities
- The generation of manufacturing instructions at different levels of detail, including routes, selection of machining processes, cutting speeds, feed rates and surface finish, and NC programs

The use of rule based techniques for the representation and use of process planning knowledge requires:

- A decomposition of process planning tasks into manageable subtasks, using say input–output analysis and algorithmic or rule based solutions, which is a difficult task
- Facilities for formulating and inputting grammar based rules
- Mechanisms for the application and control of rule sets, which is also a difficult task

A typical example of an expert system for application in manufacturing, called EXCAP (EXpert Computer Aided Process planning system), capable of planning a sequence of machining operations, with machines, clamping arrangements and tooling determined automatically, is discussed in Reference [24c]. A complete product model, including simple geometric and functional information and tolerances, is proposed. In the planning mechanism, a 'tree' of machining sequences is developed by identifying volumes of metal which must be added to the finished component to restore the original blank configuration. The patterns required for each operation, and information such as required tooling, are stored as Prolog structures, and form part of the knowledge base.

The concept of defining planning of all possible operation sequences as the inverse operation to machining was first suggested by Barash [170, 171] and later developed and extended by Mutsushima [174]. The technique involves the recursive construction of a state space, which can be used to generate a complete operation sequence including planning of individual machines, tools and cuts necessary to manufacture individual features of a workpiece. The method is also considered by Stewart [74c] for Generative Process Planning, in which knowledge within the expert system is structured and partitioned within a hierarchy, as illustrated in Fig. 7.21.

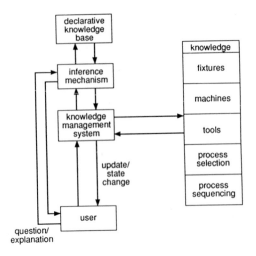

Fig. 7.21   Hierarchical knowledge management

An expert system has been developed as a planning assistant, using condition–action rules [74g]. The control program will perform such tasks as: starting machine tools, initiating part programs, monitoring system states, and scheduling preventative maintenance. The supervisory program is written in a rule based, non-procedural language, and statements are triggered on an 'IF-THEN' basis. The right hand side of the expression is executed as a set of sequential actions if the left hand side is true. For example:

IF (machining conditions are satisfied) THEN (start machining)

The expression may also be written in terms of a goal (start machining) and conditions required to satisfy the goal, in the form

(start machining) IF (machining conditions are satisfied)

This may be further extended, in the repeated goal-condition format:

(machining conditions are satisfied)
IF (part is in place)
AND (correct tools are loaded)
AND (part program is present)

This may be further repeated with a transfer of the goal to condition rules. A hierarchy of these concepts is illustrated in Fig. 7.22 (after [74g]).

Knowledge based systems also have an important role in manufacturing cell control [24a]. They will need to reason and detect space/time positions with respect to a complex 3-dimensional world model using vision sensors, and to reconfigure the cell elements, including machines, handling devices, jig and fixture elements and sensors, according to the process plan and to enable the system to recover from error conditions. These control actions could be effected within a state driven machine control system, and incorporate production rules for cell control programming using an IF–THEN state transition table.

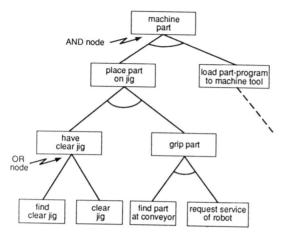

Fig. 7.22 Expert system command hierarchy

Knowledge based systems will also be required to search for collision-free paths, and to avoid stock build up, queueing and bottlenecks. At a higher level they will be required to bridge the time gap between the large planning horizons of logistic systems, for example COPICS/MRP II, and real time conditions on the shop floor. Developments are required particularly to enable systems to behave reactively; that is to base their decisions on their own internal state, the state of the environment and their set goals, and to adapt to changing conditions and system errors.

The Intelligent Scheduling and Information System (ISIS) [175, 24a] provides an interesting example of an advanced expert system which has made a significant contribution to job shop scheduling. Heuristic search in a constraint environment is used to achieve the objectives of due dates and profits. The constraints consist of: (i) organisational goals, relating for example to process inventory, level of resources, production level etc., (ii) physical constraints of machines and equipment and (iii) precedence relations and resource equipments.

Traditional 'top-down' approaches, such as MRP (see Section 8.1), lack the responsiveness required for dynamic scheduling at the shop floor level, and existing approaches cannot cope adequately with the dynamic assignment of resources at the shop floor (or work centre) level [176]. These problems have been considered in the CAM–I's Intelligent Manufacturing Management (IMM) program, in which a conceptual information flow model for a four-level factory architecture has been developed, as depicted in Fig. 7.23. Fig. 7.24 illustrates the integration of various levels of control support, including the MADEMA (Manufacturing Decision Making) concept. One of the control functions at the work centre level is to assign the machine, process and people resources, while the knowledge based MADEMA system allocates resources on the shop floor without user intervention, and provides quick response to unpredictable changes in the factory environment. The approach integrates process planning and scheduling activities in a cohesive manner and represents

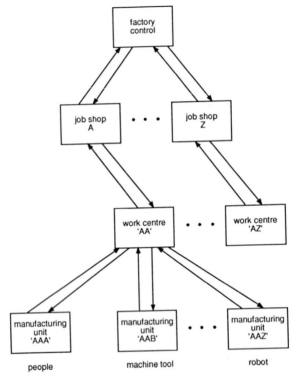

Fig. 7.23 Four-level factory architecture—generic levels of factory management and control

a decisive step in the direction of CIM. The overall knowledge based system incorporating timed dynamic events can provide a basis for dynamic intelligent scheduling and control of factory production through the job shop and down to the unit level.

7.3 Group technology

Group technology (GT) or Parts Family Manufacture is a classification method used to plan factory layout and organisation on a part family and product line basis, and the basic logic can be associated with the design of flexible manufacturing systems. It is recognised as an efficient method for rationalising small and medium batch or job shop production, according to groups and families formed by arranging the parts spectrum and the manufacturing processes into cells according to design and machining similarities. The aim is to identify common manufacturing processes and to form mutually separable clusters of machine cells and part families with similar routes, and to schedule parts for maximum machine utilisation, by emulating a flow line production process. Experience shows that the implementation of GT can reduce material

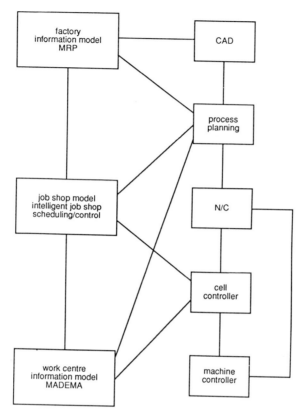

Fig. 7.24 Management control integration

movement, throughput time, set-up times, work-in-progress and inventory buffers, and simplify production control and the rationalisation of the design and planning processes. A relatively large number of GT coding systems are in use in industry, although the problems for large scale systems are computationally complex and expert system-based cluster identification algorithms may be required [177].

Geometrical information such as size and shape and machining characteristics are used to establish the families, and cluster analysis can be used to create the groups and structure the product flow layout into cells. The parts can then be produced efficiently within dedicated cells, and will not be subject to the circuitous flow paths often encountered within machine shops. However, consideration may still need to be given to cell balancing with fluctuations in product mix.

The concepts of group technology can extend beyond the limited scope of machining parts, to include other related manufacturing functions, such as CAD parts design and standardisation, process planning, production control, work measurement and social effects such as job enrichment. The concepts will

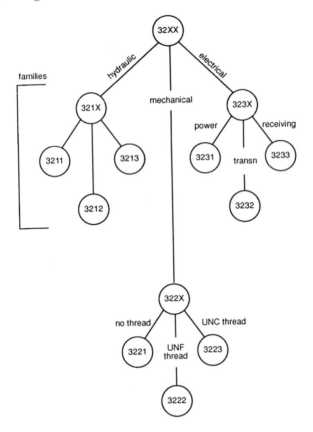

Fig. 7.25   Hierarchical coding structure

have increasing applications in hierarchical control and multistation manufacturing systems, but significant effort is required to balance the cells and to coordinate the layout structure with the data management system and the scheduling and material requirements' planning routines to achieve full benefits. The implementation of GT will also often require the coding of thousands of parts, by shape, dimension, tolerance, surface finish, production requirements etc., which can be a difficult task.

For references to GT, see References [10, 20, 25, 28, 53b, 178].

7.3.1 Parts description

A classification and coding system describes the characteristics of the part in terms of geometrical shape and/or processing characteristics. The structures used in GT coding systems include: hierarchical or monocode, chain (matrix) or polycode, and hybrid. Fig. 7.25 (after [3]) illustrates an example of a hierarchical structure, in which each code number is qualified by the preceding characteristics. This type of structure has the advantage of facilitating the representation of a relatively large amount of information with few code positions. Its disadvantages include the possible complexity of the coding system with large numbers of branches in the hierarchy.

A method of grouping which avoids hierarchical classification, based on the dynamic cluster principle, is proposed in Reference [53*b*]. Grouping has often been based on heuristic selection and classification with minimisation of an index function, and has not emphasised manufacturing constraints, as in the proposed method, which takes account of machine capacity and loading and part machining times.

Various classification and coding systems have been developed, including VUOSO (Czechoslovakia), Opitz (Aachen, West Germany), Brish (UK), K–1,2 (Japan), TEKLA (Norway) etc. The Aachen system represents a hybrid structure of chain-hierarchical form, and is probably the best known coding system [3]. The system structure provides the form code, which is expressed as:

- 1st digit: general classification of parts
- 2nd–5th digits: detailed shape of parts and machine processing
- 6th digit: supplementary code expressing dimensions
- 7th digit: material
- 8th digit: original shape of raw materials
- 9th digit: accuracy

7.3.1.1 *Parts description by topology.* Topology has been used as a basis for parts description [70], with a product viewed as a 3-complex composed of point, line, surface and volume elements or cells as subcomplexes.

An example of a rotational symmetrical part and a corresponding topological model as a 2-complex made up of vertices (0-cells), edges (1-cells) and area elements (2-cells), is illustrated in Fig. 7.26 (after [28]). The elements $F_1 = [A_1, A_2, A_5]$ and $F_2 = [A_3, A_4, A_6, A_7]$ construct the external surfaces, and $F_3 = [A_7]$ produces the internal hollow surface, where $A_i (i = 1, 2, \ldots, 7]$ are 2-form area elements.

Incidence matrices defining connections between vertices and edges (I_1) and between edges and area elements (I_2) can be used to define the basic structure of the model. The detailed shape of the part can then be determined by assigning cell values to each 2-cell, with the dimensions given as cell-coboundaries. For the model of Fig. 7.26, the standard bases are:

$$\text{Vertices: } b_s^0 = \{p_1, p_2, \ldots p_{14}\}$$

$$\text{Edges } : b_s^1 = [l_1, l_2, \ldots l_{20}]$$

$$\text{Areas } : b_s^2 = \{A_1, A_2, \ldots A_7\}$$

The incidence matrix connecting points and edges is

$$\begin{matrix} l_j \\ I_s^1 = p_i[I_{ij}], \, i = 1, 2, \ldots 14; j = 1, 2, \ldots 20 \end{matrix}$$

where $I_{ij}(0, 1)$ identifies the signed line-point connections (+ ve approaching, − ve leaving). The incidence matrix connecting edges and area elements is

$$\begin{matrix} A_j \\ I_s^2 = l_i[I_{ij}], \, i = 1, 2, \ldots 20; j = 1, 2, \ldots 7 \end{matrix}$$

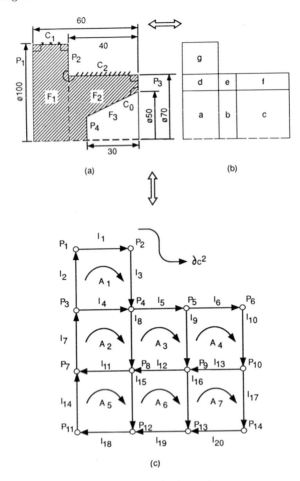

(a)   part with rotational symmetry

(b) and (c)   topological modes

Fig. 7.26   Topological model

where $I_{ij}(0, 1)$ identifies signed line-area connections, using the sign convention indicated in Fig. 7.26c.

The total part shape is represented using the 3-form elements, F_1, F_2, F_3, as:

$$C^2 = F_1 + F_2 - F_3 = b^2 \xi^2$$

where b^2 is the basis: $b^2 = [F_1, F_2, F_3]$ and $\xi^2 = [1, 1, -1]^T$

Area and form elements are connected by the basis transformation matrix

$$F_j$$
$$T^2 = A_i[T_{ij}], i = 1, 2, \ldots 7; j = 1, 2, 3$$

where the positive elements $T_{ij}(0, 1)$ identify the areas contained in the form elements. Then

$$C^2 = b_s^2 T^2 \xi^2 = b_s^2 \xi_s^2 = \sum_{i=1}^{6} A_i$$

which represents the part as a 2-chain. The surfaces of the part can then be determined by the boundary of the 2-chain, given by

$$\partial C^2 = b_s^1 I_s^2 \xi_s^2 = l_1 + l_3 + \dots$$

representing the external connection of links as indicated in Fig. 7.26c.

Part dimensions are represented by cochains. The 1-cochain, representing the signed link dimensions, is given by

$$C^{1'} = l_1(20, 0) + l_2(0, 15) + l_3(0, -15) + \dots + l_{20}(-30, 0)$$

The 0-cochain, obtained from the point co-ordinates, is given by

$$C_0' = p_1(0, 50) + p_2(20, 50) + p_3(0, 35) + \dots + p_{14}(60, 0)$$

From a topological point of view, a part can thus be completely represented by the incidence matrices, I_s^1 and I_s^2, and the 0-chain $C^{0'}$ as a bounding cocycle, which is a subgroup of group of 1-cochains.

7.3.1.2 *Parts description by graph theory*   A part can be described graphically by denoting surface elements by vertices, and the linking of adjacent surfaces by edges [28]. Each edge is assigned a value indicating the intersection angle between the two adjacent surface elements connected to that edge. Technological features such as screw, gear, spline etc. are represented by attaching a symbol indicating the feature to the corresponding surface element with a broken line.

An example illustrating the application of graph theory to the part of Fig. 7.26a is shown in Fig. 7.27, where C_i, P_i represent cylindrical and plane surface elements, respectively. The symbols S, R_0 represent screwed and knurled surface elements respectively, and the internal surface of F_2 has a taper of 20°. These properties can similarly be represented by an incidence matrix, linking the surface elements C_i–P_i.

Data items (piece parts and raw materials) and product structure data, are required to describe an assembled product. These can be represented diagrammatically, as indicated by the example in Fig. 7.28, using a tree, an incidence

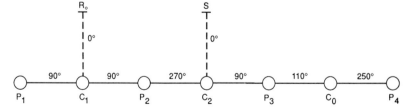

Fig. 7.27   Graph representation of part description

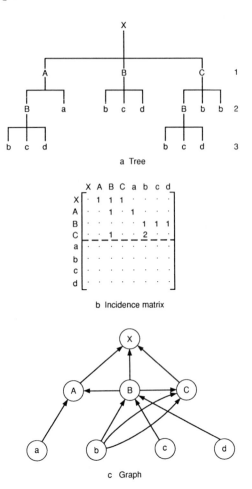

Fig. 7.28 Diagrammatic representation of product structure

matrix, and a graph structure. A bill of material or product summary list and inventory list can then be drawn up for a particular production period, to indicate the detailed make-up of the end product and quantity of part items required.

7.3.2 Lot sizing problem

The concept of grouping or class partitioning plays a central role in production planning systems, from the relatively long range planning of cell grouping to the shorter range planning of lot sizes for product mix. The number of possible solutions to the lot sizing problem will be large, which will usually preclude the implementation of any effective optimisation strategy, and a heuristic search based on simple decision rules has been suggested, using the programming language APL [70*b*].

The set of parts to be manufactured within a cell in the coming period (e.g. a week) is considered to constitute a lot. A product mix with a maximum number of different parts will be required to balance the load on the machine tools and to achieve a shortest possible lead time. On the other hand, the product mix will be restricted by the tool magazine capacity of each machine tool.

A tertiary array and Boolean transformation matrices can be used to characterise the lot sizing problem in terms of the basic entities representing parts, machine tools, number of tools and lead times. The formulation also includes an objective function expressing the total lead time of all lots, to be minimised subject to the tool constraints. A complete exhaustive search of the resulting combinatorial problem will be prohibitive, and selection criteria based on a heuristic strategy involving priority numbers has been proposed. This aims to ensure that the key machine tool, which primarily determines lot sizing because of tool constraints, is kept permanently busy as the critical (bottleneck) machine.

Chapter 8
Management Control Aids

8.1 Introduction

Many OR techniques developed since the early 1940s were applied to manufacturing problems, and particularly to the ordering and batching of components, and work scheduling for the m-machine, n-job problem [178]. The solutions did not attempt to integrate company activities, and were impractical for large and realistic values of m and n.

An integrated approach to production control was being considered during the 1960s, and Plossl and Wight [179] and others considered the need to integrate individual production processes within a total production control system. The introduction of Orlicky's dependent demand principle [180] during the mid-1970s provided an improved ordering technique which evolved into the conceptually simple technique of Material Requirements Planning (MRP), with an emphasis on the total system and current and future item requirements [181].

During the 1970s, principles of scientific management were used to develop mathematical techniques for solving problems such as demand forecasting, inventory planning and control, and shopfloor scheduling. The systems as implemented gave no information on how particular results were obtained, and they generally failed to be accepted or extended to larger scale problems because of a lack of understanding and the computational problems involved.

In MRP, inventory control is based on available capacity and on-hand inventory, to meet a forecasted future demand. The process generates a master production schedule (MPS) which indicates product quantities up to a year in advance. This is broken down into specific raw materials and the phasing of components to be ordered and manufactured in the Bill of Materials that defines the product, obtained by back tracking on delivery dates and taking account of lead times and in-house inventory. The objectives of an MRP system are thus to identify order quantities for every item and timing, and to schedule delivery, and thus to provide an effective method for material purchasing and for the planning and control of inventories.

MRP conceptually operates as an 'order inlet control', 'push' or 'top-down' system, unlike the Just-in-Time (JIT) or Kanban approach acting as a 'pull' or 'bottom-up' system, depicted in Fig. 8.1. It lacks feedback control from the factory floor required to avoid physical delays and leads to increasing levels of work-in-progress and shortages, which can only be counteracted after some delay with the next MRP run [17c, 79]. It is also inflexible and unable to deal

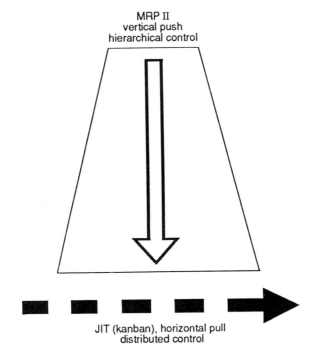

MRP II
vertical push
hierarchical control

JIT (kanban), horizontal pull
distributed control

Fig. 8.1 Push (top down)–pull (bottom-up) strategies

with unexpected events in manufacturing.

Another major difficulty with MRP is the lack of consideration of capacity constraints and the need to forecast batch sizes and lead times, and in many cases it is an inappropriate technology for scheduling since it does not address the overall requirements of the business [20]. Alternative systems of materials control, such as the Kanban system, are also less dependent on forecasting.

8.2 MRP II

The Manufacturing Resource Planning method, MRP II, developed in the USA as an extension of MRP, incorporates other aspects of the company's business including sales, purchasing and finance, and capacity requirements planning (CRP). It provides a management aid with closed loop feedback from the operational level [7]. Information such as the late delivery of raw material can, for example, be used to activate a delay in a previously scheduled work order with consequent changes in capacity planning, master production scheduling, production planning and material requirements planning, and a rescheduling of orders to other suppliers.

The usual MRP II production planning strategy is illustrated in Fig. 8.2 (after [157]). The system produces a master production plan which activates MRP and capacity requirements planning. MRP specifies the quantity and

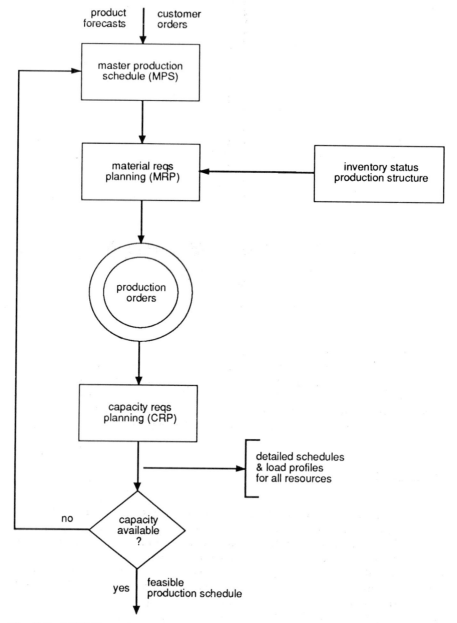

Fig. 8.2   MRP II production strategy

quality of the different types of material required with sufficient lead time for the manufacture of a range of products and components by due dates. A more detailed representation of the activities involved in a closed loop MRP II system is illustrated in Fig. 8.3 (after [20]).

MRP II provides a 'top down' approach to integrated closed loop production control, which can lead to reductions in inventory and purchasing costs, improved productivity and improved customer service. The concepts introduce pro-active rather than reactive decision making under changing conditions, and fulfil all the requirements for an integrated advanced factory management system. Conventional packages, however, are usually based on infinite capacity scheduling which may result in overload of the production stages [17*d*].

Most production management software is today based on MRP II principles, and successful implementations in some large companies have produced significant results, in reducing inventories and work-in progress. Large MRP II systems, however, are complex and expensive, and usually include a relatively large number of software modules in relation to those which are often required. They can take long periods of time to implement, and often do not realise their full potential through lack of understanding. Few companies apparently have implemented MRP II systems successfully, and effectively closed the loop between the business plan, and production and material plans, through the use of Master Production Scheduling, although they presently provide the only computer aided support required for large companies to carry out long term planning [7]. The advantages of the product-structure-based MRP II system are in global planning and those of the Just-in-Time system (Section 8.4) in short term execution, and the two approaches should be viewed as complementary and possibly combined to form a mixed strategy.

8.3 Optimal production technology (OPT)

The difficulties of implementing MRP II have stimulated many manufacturing companies to turn to simpler but more radical production support systems, such as Just-in-Time, implemented successfully in Japan, and the proprietary system Optimal Production Technology, originated by Eliyahu Goldratt [8]. The innovative technique (OPT) is a push-type, process-structure-based procedure, and corrects the difficulties of previous methods based on infinite loading procedures which cannot plan priorities and capacities simultaneously [90, 178]. OPT adopts a two stage approach for generating schedules—an infinite loading routine identifies critical (bottleneck) work centres, and the OPT finite loading procedure is used to develop feasible schedules for the critical resources. OPT looks at all the constraints simultaneously, as an integrated planning and scheduling procedure, and a master schedule results from the schedule at the bottleneck stage [157]. Following the best scheduling of resources at the bottleneck stage, operations at the remaining stages can be scheduled forward and backward in time.

OPT thus effectively focuses on the identification of bottlenecks, caused for example by a slow (but critical) machine or the late supply of a part, and produces an optimal reschedule to reduce lead time. OPT systems have been

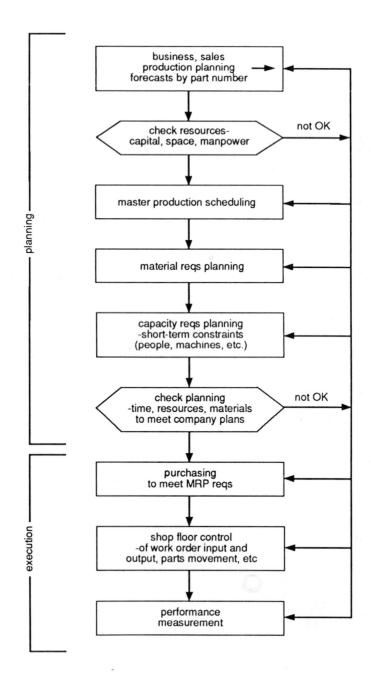

Fig. 8.3   Model of a closed loop MRP II system

implemented successfully in many large companies worldwide and, although expensive, they have had a very high success rate. They can operate as the enabling technology for JIT and as a complementary tool for use with MRP systems. There is now, however, a tendency to change production strategies from the top-down planning approach using MRP II, and bottom-up methods such as JIT, to OPT, to enable bottlenecks to be shifted throughout the whole factory [182].

8.4 Just-in-Time (JIT)

The JIT philosophy developed from attempts by the Toyota company of Japan to reduce run lengths, changeover times and components stocks, during the period 1930–60. It is based on the concept of make-to-order, and not make-to-stock, often found in Western manufacturing practice, and is now used successfully in many Japanese industries. It has also been adopted successfully by manufacturing companies in the USA and Europe since the mid-1980s. Inventory, materials handling, work-in-progress, space and setting-up time, are all categorised as 'waste' which must be reduced or eliminated, and bottlenecks 'cannot' be tolerated. It aims, broadly, to reduce lead times and inventory levels to a minimum, and improve quality levels and customer service.

JIT considers that the cost of holding and managing stock can be eliminated if items are made or ordered only when required. Removal of the protection and security of inventory, buffer stocks and long lead times then increases the risk of interrupted production and demands a fundamental reorganisation of the suppliers of raw material and components. They will usually be required to be close to the factory, and able to operate tightly controlled delivery schedules by delivering just the right item, at the right place and right time. The risks must also be mitigated by tight control of all manufacturing activities, including maintenance, scheduling (based simply on demand), quality control and reliability, and this requires continuous attention to the reduction of machine set-up times, lead times, internal transportation costs and overheads. Reduction of changeover times enables batch sizes to be reduced, which improves material flow and reduces inventories.

JIT also requires consideration of other issues, such as:

- Logistics, to reorganise, for example, the flow of parts and products through simple unidirectional routes.
- Selection of products compatible with material flow control.
- An effective planning and scheduling system for tools.
- Reorganisation of production areas with relocation of transportation systems, buffer stocks, and decentralisation of stock organisation.

The JIT approach is not an expensive or sophisticated computer based technology, and can produce considerable rewards when applied successfully. It is based on a market demand or 'pull' regime, and introduces a movement towards the concept of synchronised and continuous process manufacturing. It is best fitted to product flow with relatively long term stable demand and predictable sales, with repetitive manufacture of items such as television sets, automobiles etc.

The major problem in a pull system such as JIT is the need to plan for reduced inventory and buffer stocks, with possible short term losses traded off against the long term gains resulting from continuous flow and increased productivity. The problem, incorporating queueing interaction, requires further investigation to produce efficient algorithms, say within an optimal control framework.

JIT as operated in Japan with flexible automation represents a highly integrated production, sales and distribution system, producing continuous and efficient flow of material, components and products through the overall system. It requires the development of good forecasts and production plans, and is driven by a detailed master production schedule.

For references to JIT see References [2, 7, 25, 44, 90, 178, 183—185, 87].

8.5 Kanban

Kanban, of Japanese origin, focuses on capacity control and is a mechanism for regulating the flow of the shop floor, as required in the JIT philosophy. It is in essence, a re-order system using a job ticket (kanban) accompanying a part (or tool), which is returned to a source feeding station on completion of the work to trigger the production of a new part. Its basic concept, as with JIT production, is fast demand-activated flow of manufacturing and assembly of discrete products, operating on the basis of fixed orders, and reduced inventory, set-up times and lead times, and is mainly applicable in high volume repetitive production.

The pull-type systems with limited buffer sizes are more sensitive to cycle time variations because of the limited work-in-process, in contrast to the frequently used push systems, which can achieve a high tool utilisation at the expense of large work-in-process. Pull-type decisional architectures, although unable to handle large demand fluctuations, can be more efficient for customer-oriented production, and push-type systems, although they may not be able to reduce work-in-process, can work well for stock-oriented production. This suggests a unified integration of the different philosophies involving concepts of batch (customer-oriented) and repetitive (stock-oriented) manufacturing.

The JIT/Kanban approaches, however, may not be particularly relevant to Western practices of batch manufacturing control, where product variety and flexibility of delivery are essential for survival. Some of these problems can possibly be remedied by the integration of OPT techniques with JIT, both of which are oriented to making only what can be sold.

For references to Kanban see References [2, 5, 7, 21, 79, 90].

8.6 Quality control

Product quality and customer satisfaction must be considered as major priorities within the corporate objectives of a manufacturing company. The concept of quality must extend throughout the manufacturing process, from the design stage to final assembly, and be linked to the concepts of Inspection, Feedback and Action. A relatively large number of quality control functions will usually be required at different levels of a manufacturing process, to monitor and correct deviations from specified levels [10].

The basic concept for a 'zero defect' quality control system, advocated by Shingo [186], is based on the idea of applying control functions where the defects originate, and using complete rather than sampling inspections. At the other extreme, quality control in advanced manufacturing systems will be distributed and structured hierarchically, and integrated computer based systems may be used to co-ordinate and supervise the overall quality control function [17e]. The quality control system will also need to respond dynamically to frequently changing production runs, and research is needed to develop simplified distributed data bases, with facilities for frequent updating of files required with changing samples, test sequences and tolerances.

The dimensions of the work pieces during machining and of the finished products, and also the production and assembly processes must be monitored for quality and performance and also errors such as deviations from standard tolerances, machine and fixture misalignments, tool wear etc. The quality control system must also be planned with regard to the location of sensors and the frequency of measurements.

Efforts must be made to locate all errors and defects during manufacture and assembly, and to initiate appropriate corrective action within the production process, possibly with the quality control system acting within an adaptive control loop. An integrated quality control system planned within a company-wide systems approach has been conceived as shown in Fig. 8.4 (after [10]), in which corrective control of the planning or scheduling processes is initiated by errors between measured quality and set points determined by predefined quality specifications or standards.

The development of automated inspection and quality control systems using testing strategies, derived directly from part design data, will require a model representation of the manufacturing process, monitoring with intelligent sensors, and adaptive control for automatic adjustment of process variables. Adaptive control will operate by monitoring, identifying and controlling machine level operations, subject to uncertain disturbances produced for example by machine failures and tool wear. The automated part inspection system will model the current state of the system and predict abnormalities such as tool wear, and by communicating with software modules in the common manufacturing data base, will adapt to achieve the optimal part design. The machining process model will, in general, be highly nonlinear, and advanced nonlinear identification techniques may be required for parameter estimation.

Other fully automated inspection systems are being developed which derive and execute testing strategies directly from part design data [9b]. They incorporate a 3-dimensional mechanical co-ordinate measuring machine with touch trigger probes and optical sensing, and a hierarchical control system consisting of state tables and data structures representing multiple-state machine modules. Machine accuracy enhancements can be considered based on static positioning errors, thermal errors, tool chatter and machine vibration. A system for the automatic detection of decayed areas of cutting tools by image

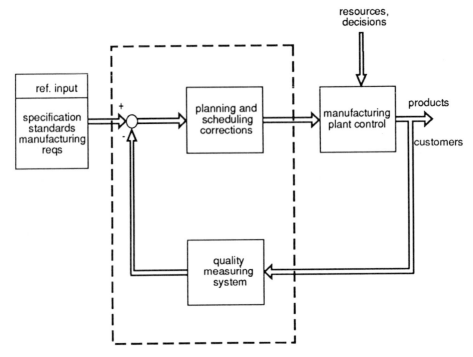

Fig. 8.4   Integrated quality control

processing a sequence of TV images is also being considered to predict tool life and support a replacement policy, using a radial Fourier descriptor to discriminate between decayed and non-decayed areas.

8.6.1 Total quality management

Total Quality Management is the philosophy that continual improvement in the quality of all activities involved in manufacturing will produce better product quality. It requires team activity for investigating sources and means of eradicating problems and errors, and must be encouraged and co-ordinated by top level management. This produces an improved working environment which cultivates co-operation at all levels and leads to improved products.

The implementation of an efficient scheme of Total Quality Management will require an effective in-house programme of education and training to instil the correct procedures and a work philosophy. It will also be necessary for higher levels of management to move into more active service or operational roles, which will facilitate the 'on-line' implementation of new developments and ideas. The total philosophy will incorporate the traditional methods of product quality assurance, and ensure that they are kept in perspective and do not inhibit the wider developments.

Chapter 9
Education and training

9.1 Introduction

Manufacturing Systems Engineering is a broad based multidisciplinary subject embracing many branches of engineering such as mechanical, electronic, electrical and control engineering, as well as computer science, and economics and business studies. It involves a wide range of activities concerned with planning, designing, constructing and operating large complex systems, and the manufacturing environment provides many exciting and intellectual challenges.

Industry, however, is facing severe difficulties in recruiting staff with the wide-ranging skills needed to contribute effectively in the manufacturing systems field. A major constraint in exploiting the gains to be derived from automated manufacturing is not the shortage of technology, but rather the manpower with the skills and expertise needed to design and implement advanced integrated manufacturing systems. The complexity of new technology, and the pace with which it is being introduced to support the requirements for good design and efficient manufacture, is increasing the demand for new approaches to the education and training of people to enable them to implement advanced technology. A major education job exists to provide courses involving all aspects of integrated manufacturing systems engineering for the retraining of existing staff and for new engineering and science graduates.

Broadly based project-oriented training courses incorporating multi-disciplinary team work must be developed, and directed towards industrial processes such as discrete part manufacturing systems which require the integration of a wide range of methodologies, technologies, processes and materials.

A new breed of engineer is needed, possessing multi-disciplinary engineering, business and management skills [187]. This will require training to be focused on the systems approach to manufacturing problems, with an emphasis on digital electronics, computer programming and software engineering, systems engineering, knowledge based systems, multi-criteria decision making and management control.

Many manufacturing companies lack staff with the experience and abilities necessary to design, implement and operate complex, large scale integrated manufacturing systems, and existing staff will often have great difficulty in handling the cognitive and motivational problems emerging from the new technologies. Greater emphasis may thus need to be placed on the introduction of decentralised approaches to smaller scale automated systems before investment in fully integrated processes.

The structure of a possible set of modules suitable for retraining or for forming the basis of a Master's degree course in Manufacturing Systems is outlined. The aim would be to provide students with a range of systems concepts and techniques which would enable them to appreciate and understand the complex interrelationships existing in integrated manufacturing systems. The course overall would emphasise the computer integration and structured analysis of manufacturing within a framework based on corporate strategies, manufacturing policies and business and social issues.

9.2 Proposed Course Modules

INTEGRATED MANUFACTURING SYSTEMS

First Term

1 *Manufacturing systems*
Overview, hierarchical structures, systems concepts, planning, scheduling, control. Structured analysis. Modelling, discrete event simulation, applications

2 *Strategic planning*
Manufacturing organisation, strategy. Marketing, sales, forecasting. Project management, financial control, accounting. Management and financial modelling. Decision support systems

3 *CADCAM*
Product design for manufacture, functional characteristics, manufacturing constraints, material selection. Graphical commands, data bases, postprocessors, IGES. Software packages. Machine tool programming. Knowledge based systems

4 *Computer hardware—Software engineering*
Computer architectures, operating systems, networks, Open Systems Interconnection, MAP, TOP protocols. Man–computer interaction. Selection of computer systems, interfaces. Knowledge based systems. Software design tools, programming methodology, testing.

5 *Case studies—Group projects*
To investigate the activities involved in the design and implementation of an advanced manufacturing system required to produce a range of products, in collaboration with industry
Tutorials—Group discussions with video support

Second Term

6 *Flexible manufacturing systems I*
Cell, workstation control, production planning, scheduling, machine tool monitoring, sensing devices, quality management and control. Tool management systems.

7 *Flexible manufacturing systems II*
Production planning and control. Management information and control systems, MRP II, OPT, JIT, Kanban, CAPP, logistics, inventory control, group technology. Knowledge based systems.

8 *Robotics*
Kinematics, programming languages. Path planning, trajectory control. Sensors, vision systems. Software safety systems. Materials handling, AGV,

storage, automated assembly. Modelling, simulation. Economics. Intelligent robots.

9 *Computer integrated manufacturing*

CIM strategy. Integration of design, planning, scheduling, control, quality assessment and corporate strategy. Design to manufacture concept. Human factors, organisation design. Developing a CIM strategy. Support funding.

Third Term

Main Course project

An in-depth study of a specific problem relating to the design, implementation and operation of an Integrated Manufacturing System within an actual production environment.

Chapter 10
Conclusions

Future advances in manufacturing depend critically on the development of real time facilities incorporating distributed data bases and communications technology, and integration through CADCAM and computer aided production planning combined with the concepts of OPT, JIT and group technology. Integration of design and operational planning and control offers major advantages through improved productivity and customer satisfaction, although the availability of a totally integrated CIM package handling all activities within a manufacturing company is unlikely. The economic benefits of complete integration are difficult to quantify and few companies have yet bridged the gap between MRP II and the design and manufacturing activities. Future initiatives will need to develop towards CIM in a planned and orderly way, adopting steady incremental improvements and attention to detail, as with the Japanese philosophy, rather than with sudden change.

Major challenges lie in the application of knowledge based systems to the design process, for multi-resource scheduling with moving horizons and uncertainty, and to discrete event modelling and simulation. Further developments are also required in computer controlled assembly incorporating 3-D computer vision systems, and in the implementation of signal processing techniques for real-time analysis of process variables using advanced sensors to identify unsafe working conditions. Further study of the role of rule-based, discrete event state models is required for defining reference levels and goals at the factory, shop, cell and workstation levels for decision making in the integrated design–manufacturing system. The optimal selection and design of cost-effective storage buffer capacity in the manufacturing flow process is an important requirement which demands further detailed study. Improved methods for locating and tracking parts through the production process, and for communicating instructions from the design office to shop floor machines, are also required [7].

Large interconnected manufacturing systems will usually be connected in subordinate ways and linked with a hierarchy of computers, and they cannot effectively be analysed and controlled using existing mathematical tools. Research, encompassing system representations, structural modelling and system planning, scheduling and control subject to environmental constraints and uncertainty, and the evaluation of multiple performance criteria with conflicting objectives and trade-offs involving efficient production, work-in-process, buffer storage, product quality and customer satisfaction, is urgently required to support the development of improved methodologies. There is a

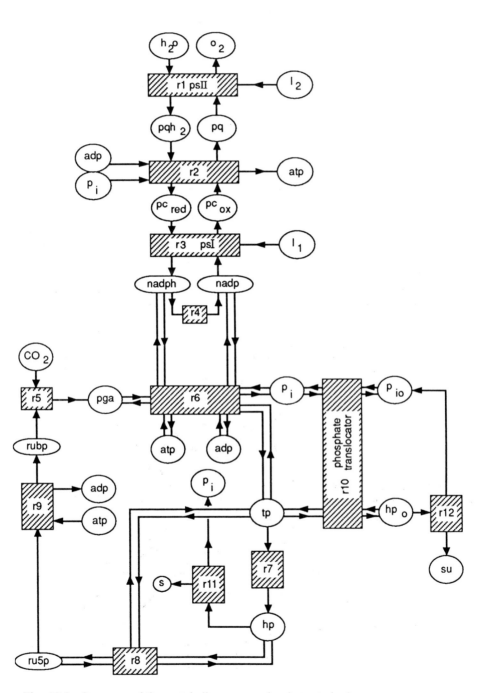

Fig. 10.1 Structure of the metabolic processes in photosynthesis

need to return to the fundamentals of manufacturing system planning and control, and to raise the intellectual profile of the complex problems involved.

Future factories will move to centres of demand to reduce rising distribution costs, and some large scale manufacturing will diversify into smaller companies focusing on specialised production needs. A move from the machining and processing of metallic materials to the forming and welding of composite products and assemblies in fibre reinforced plastics is expected, which may change the type of production system required in 10—15 years time. Metal forming involving stamping, cutting, welding, glueing etc. will also require expansion and automation, and present flexible manufacturing systems may be limited in their abilities for these applications.

The functions involved in integrated production and business systems have certain resemblances to those of a biological system with material inputs and output products and similar hierarchical flows of information, and with multilevel control used to activate pre-programmed control at the lower levels [9c, 188]. The structure of the metabolic processes involved, particularly in photosynthesis, which demonstrate the similarities involved, is illustrated in Fig. 10.1 (after [189]). The analogy could be further strengthened and highlighted by the development of a Petri net model for this system, with input places representing reacting chemicals, transitions representing reactions, output places the results of a reaction, and tokens the number of molecules

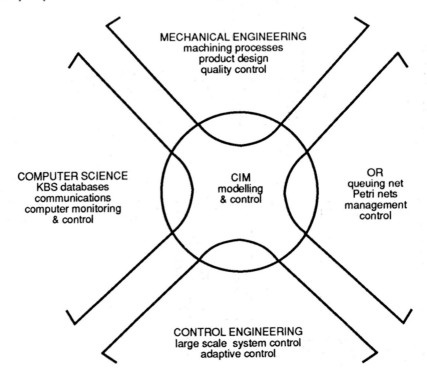

Fig. 10.2 Integration of disciplines for CIM modelling and control

involved [164], although this might be a difficult task. Neural networks could also have possible applications for modelling and decision making in manufacturing systems.

Developments in automated manufacturing are having far-reaching effects, particularly in education and training and in the demands for skilled manpower with wide ranging talents and experience. There must be an integration of the different disciplines involved in manufacturing, which are not presently structured correctly for this activity, and Fig. 10.2 indicates the overlapping component disciplines which will be needed to provide a framework for the future CIM modelling environment. The manufacturing field will provide enormous opportunities for interdisciplinary research, and many of the fundamental concepts originating particularly from the Control Systems field [190] will provide a unifying basis for future theoretical developments. However, although the concepts are directly relevant, their applications in manufacturing will be different from their realisations in traditional control applications.

Petri net modelling

Basic definitions [161, 162]

A Petri net consists of a finite set of places P, a finite set of transitions T, and input (I) and output (O) functions relating transitions and places.
A Pn *structure* C, is a four-tuple

$$C = (P, T, I, O), P = \{p_1, p_2, \ldots, p_m\}, T = \{t_1, t_2, \ldots, t_n\},$$

$$n, m \geq 0$$

An example of a Pn structure is stated as follows:

$$C = (P, T, I, O), P = \{p_1, p_2, p_3, p_4, p_5, p_6\}, T = [t_1, t_2, t_3, t_4, t_5\}$$

$$I(t_1) = \{p_1, p_4\} \qquad O(t_1) = \{p_3\}$$

$$I(t_2) = \{p_2\} \qquad O(t_2) = \{p_1, p_4\}$$

$$I(t_3) = \{p_3\} \qquad O(t_3) = \{p_1, p_4\}$$

The structure can similarly be defined in the form

$$I(p_j) = \{t_i, \ldots\}, O(p_k) = \{t_j, \ldots\}$$

A Pn *graph* represents the Pn structure as a bipartite directed multigraph, with, in general, multiple arcs connecting the places and transitions. Fig. A1.1 (after [191]) illustrates the equivalent Pn graph for the above Pn structure. The dual of an unmarked Pn, $C = (P, T, I, O)$ is the Pn, $\bar{C} = (T, P, I, O)$, with places and transitions interchanged in the same structure. Duality is an interesting concept, although it has not been exploited in Pn theory, and is difficult to define for a token-marked Pn.

The execution of a Pn is controlled by the position and movement of markers (tokens), the numbers and positions of which may change during an operation. The tokens are represented by small dots, as shown in Fig. A1.1. The marking can be defined by the n-vector $\mu = (\mu_1, \mu_2, \ldots, \mu_n)$, where μ_i indicates the number of tokens in place p_i. A marked Pn is sometimes denoted $M = (C, \mu) = (P, T, I, O, \mu)$.

Execution rules

A Pn operates by firing transitions which remove tokens from the input places and distribute new tokens to the output places. A transition may fire if it is enabled, which requires at least one token in each of its input places. The firing of a transition removes all of the enabling tokens from its input places and

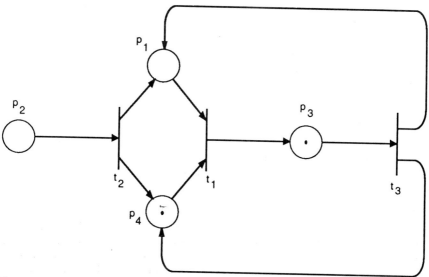

Fig. A1.1 Initial Petri net representation

deposits a token into each output place for each arc to the place. Multiple tokens are produced for multiple output arcs. A transition t_1 with $I(t_1) = \{p_1\}$ and $O(t_1) = \{p_2, p_3\}$ is enabled with one token in p_1, and fires by removing a token from p_1 and depositing one token in p_2 and one in p_3. Transition firings can continue if there exist at least one enabled transition.

The *state* of a Pn is defined by its marking, and the change in state caused by firing a transition is given by $\delta(\mu, t_j) = \mu'$ if t_j is enabled, where δ is the change function, and μ' is the new marking resulting from removing tokens from the inputs of t_j and adding tokens to the outputs of t_j.

The *reachability set* $R(C, \mu)$ of a Pn C with marking μ is defined to be the set of all states or markings which are reachable from μ. For example, consider the Pn of Fig. A1.1, which has the initial marking (state), $M_0 = (0011)^T$. Transition t_2 fires by removing a token from p_2 and adding tokens to p_1 and p_4, giving the state $M_1 = (1002)^T$ shown in Fig. A1.2. Transition t_1 can then fire and transform state M_1 to state $M_2 = (0011)^T$. In the general case, a state M_n is said to be reachable from a state M_0 if a legal firing sequence exists.

If two or more transitions have a common input place they are said to be in *conflict*, and firing must be selected using a priority protocol. For example, in the Pn structure $P = \{p_1, p_2, p_3\}$, $T = \{t_1, t_2\}$ with

$$I(t_1) = \{p_1\} \qquad O(t_1) = \{p_3\}$$
$$I(t_2) = \{p_1, p_2\} \qquad O(t_2) = \{p_3\}$$

with marking $M_0 = (110)^T$, the transitions t_1, t_2 are in conflict and only one transition can fire at a time.

If each place in a Pn has single incoming and outgoing arcs, it can be drawn as a directed graph (digraph), with vertices corresponding to transitions and arcs to places. This form of Pn is known as a *marked* graph, which can represent concurrency but not conflict and is a structurally restricted subclass of Pn. Finite state machines are similarly restricted Pns with each transition having one input and one output. A *free choice* Pn is a structure where every place p is either the only input place of a transition or there is only one transition which has p as an input place [164].

Pn state equations

Pns have been studied using graph theory, linear algebra and the concept of discrete-time state equations [191—193]. The incidence matrix of a Pn is an $n \times m$ matrix of integers, $A = [a_{ij}]$, with m places and n transitions. The entries are given by

$$a_{ij} = a_{ij}^+ - a_{ij}^-$$

where a_{ij}^+ is the number of arcs from transition i to the output place j, and a_{ij}^- is the number of arcs to transition i from the input place j. A marking or state vector M_k is defined as the $m \times 1$ column vector of nonnegative integers, with the jth entry denoting the number of tokens in place j immediately following the kth firing. An $n \times 1$ control vector U_k, with a unity entry in the ith position, defines the execution of the ith transition at the kth firing. The Pn state equation can then be represented in the form of a discrete-time system, as

$$M_k = M_{k-1} + A^T U_k, k = 1, 2, \ldots \tag{A1.1}$$

where M_k represents the marking resulting from the marking M_{k-1} by firing transition i, and superscript T denotes the matrix transpose. Matrix A is the

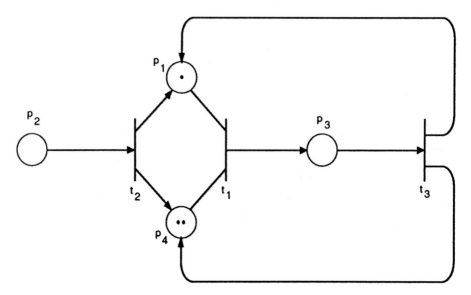

Fig. A1.2 Petri net of Fig. A1.1 after firing

transition-to-place incidence matrix, with entries $a_{ij} = 1, -1$, or 0 if transition i has an outgoing arc to place j, an incoming arc from place j, or no arc between them, respectively. The ith row of A denotes the token changes in the p places when transition i fires once. For marked graphs, A reduces to the vertex-to-arc incidence matrix of a digraph. The control vector U_k enables an appropriate column of A^T to be selected such that

$$M_{k-1} + A^T U_k \geq 0 \text{ for each } k$$

to enable M to be a vector of nonnegative integers.

Example: For the Pn of Fig. A1.1, the state M_1 resulting from initial M_0 by firing transition t_2, shown in Fig. A1.2, is given by

$$M_1 = \begin{bmatrix} 0 \\ 0 \\ 1 \\ 1 \end{bmatrix} + \begin{matrix} p_1 \\ p_2 \\ p_3 \\ p_4 \end{matrix} \overset{\begin{matrix} t_1 & t_2 & t_3 \end{matrix}}{\begin{bmatrix} -1 & 1 & 1 \\ 0 & -1 & 0 \\ 1 & 0 & -1 \\ -1 & 1 & 1 \end{bmatrix}} \begin{bmatrix} 0 \\ 0 \\ 1 \end{bmatrix} = \begin{bmatrix} 1 \\ 0 \\ 0 \\ 2 \end{bmatrix}$$

Summing eq. A1.1 for $k = 1, 2, \ldots, n$ gives

$$A^T \Sigma = \Delta M \tag{A1.2}$$

where $\Delta M = M_n - M_0$ and $\Sigma = \Sigma_{k=1}^n U_k$ is the firing count vector whose entries give the number of times each transition fires in sequencing from M_0 to M_n.

Certain structural properties of the Pn can be characterised in terms of the incidence matrix A, and particularly the controllability and reachability of Pns can be related to maximal matchings of the bipartite graph. Thus, considering eqn. A1.2 in the form

$$[A^T A^T \ldots A^T] U = \Delta M$$

a necessary condition for a Pn to be completely controllable is

$$\text{rank } A^T = p \tag{A1.3}$$

which is a result well known in control system theory.

Eq. A1.3 is a necessary and sufficient condition for complete controllability of marked graphs, enabling any initial state to reach any other state. Thus, a marked graph is completely controllable if and only if the underlying graph is a tree. The condition of eqn. A1.3 is often not satisfied in practical Pns, and the reachability problem is to find a legal firing sequence which will transform M_0 to a given state M_n.

If a given Pn is not completely controllable, with the rank $r(A) < p$, then it can be shown [191] that the necessary condition for the existence of a firing sequence which transforms an initial state M_0 to another state M_n, is the reachability condition

$$B_j \Delta M = 0 \tag{A1.4}$$

where

$$B_j = p - r[\ \overset{p-r}{I} \vdots \overset{r}{-A_{11}^T (A_{12}^T)^{-1}}]$$

with A partitioned in the form

$$A = \begin{array}{c} \begin{array}{cc} p-r & r \end{array} \\ \begin{bmatrix} A_{11} & A_{12} \\ A_{21} & A_{22} \end{bmatrix} \begin{array}{c} r \\ t-r \end{array} \end{array}$$

det $A_{12} \neq 0$, and $B_f A^T = 0$. Eq. A1.4 is a necessary and sufficient reachability condition for a marked graph.

Example: For the Pn of Fig. A1.1 and A1.2

$$A_{11} = \begin{bmatrix} -1 & 0 \\ 1 & -1 \end{bmatrix}, A_{12} = \begin{bmatrix} 1 & -1 \\ 0 & 1 \end{bmatrix}, A_{21} = [1 \quad 0], A_{22} = [-1 \quad 1]$$

$$B_f = \begin{array}{c} \begin{array}{cccc} p_1 & p_2 & p_3 & p_4 \end{array} \\ \begin{bmatrix} 1 & 0 & \vdots & 0 & -1 \\ 0 & 1 & \vdots & 1 & 1 \end{bmatrix} \end{array}, \Delta M = M_1 - M_0 = \begin{bmatrix} -1 \\ 0 \\ 1 \\ -1 \end{bmatrix}$$

and $B_f \Delta M = \begin{bmatrix} 0 \\ 0 \end{bmatrix}$, indicating that M_1 is reachable from M_0.

It is interesting to note that circuit theoretic concepts have been developed and used effectively for the study of marked graphs [193]. For the differential marked graph ΔG with the arcs and vertices represented by the corresponding entries of ΔM and Σ, respectively, Kirchhoff's voltage law $(C_c^T E = 0)$ can be identified with eqn. A1.4 $(B_f \Delta M = 0)$, and the transformation of node or tree branch voltages $(E_0' \equiv \Sigma)$ to branch voltages $(E \equiv \Delta M)$ can be identified with eqn. A1.2 $(A^T \Sigma = \Delta M$ or $A^0 E_0' = E)$. The matrices C_c^T, A^0 represent the transposed branch-mesh matrix and branch-node-pair matrix, respectively [35], which correspond with the matrices B_f and A^T respectively. The problem of finding a firing count vector Σ then corresponds analogously to solving eqn. A1.2 for the t node voltages from a set of p given branch voltages ΔM in ΔG.

This circuit analogy identifies voltage-related conditions in the marked graph, and it is dificult to interpret the existence of current flow, unless Kirchhoff's current law $((A^0)^T i = 0)$ is identified with eqn. A1.4 and the current transformation equation $(i = C_c i^{C'})$ with eqn. A1.2. In this case, ΔM would correspond to branch currents i and Σ to mesh currents $i^{C'}$. The combination of the combined voltage–current concepts appears to be difficult with discrete-event phenomena, although it could be feasible to develop techniques involving network tearing and sequential solution using a basic tree structure [35], for the restricted Pn.

The state variable formulation suggests the possibility of applying other system theoretic results, concerning for example optimal control, to Pns, using standard performance indices with time/cost assignments to transitions and places [191].

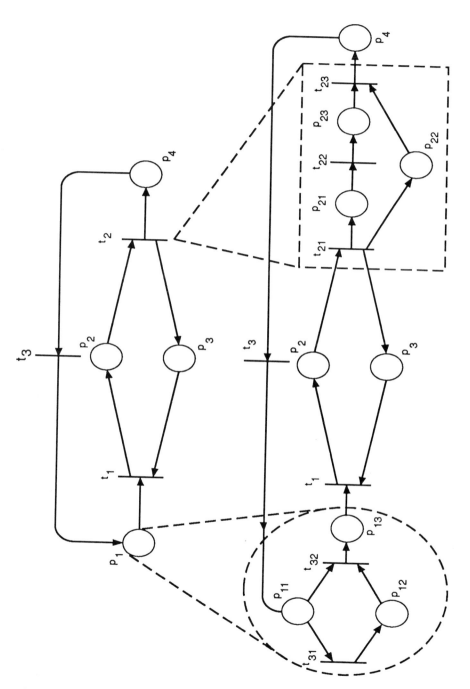

Fig. A1.3 Structured Petri net modelling

Timed Petri nets

Standard Pns can be extended by incorporating a timing mechanism or enabling times on transitions and places operating on a timed schedule [194, 195, 168]. These will indicate the duration of the event modelled by the corresponding transition, and define the time elapse before an enabled transition can fire. The timing mechanism will enable simple measures such as time spent in a marking, the average number of tokens in a place, and the relative frequency rate of transition firing to be obtained, which can be combined to provide overall performance indices such as throughput, work in progress etc.

A timing mechanism can be defined with an enabling time vector $D = (d_1, d_2, \ldots d_n)$, with entries d_j indicating the duration of the event or transition t_j. Delayed transitions are also often represented using a rectangular block with a labelled time delay, and a place with concentric circles and loading and discharging times.

Structured modelling of Petri nets

A Pn graph will often be large and complex for a practical system and may be too difficult to handle. It then becomes necessary to reduce the level of detail at an aggregate level, using say a structured approach by expanding transitions and places into detailed subnets [196, 162]. This requires conditions to be defined under which a subnet can be substituted for a single transition while preserving properties such as liveness (absence of deadlocks) and boundedness (absence of overflows). The concept of using detailed subnets to represent single places or transitions at a global abstract level is illustrated in Fig. A1.3 (after [162]).

The ESPRIT–CIM Programme

The European Programme for Research in Information Technology (ESPRIT)–CIM initiative was launched in 1984, with the long term objective of establishing methodology for implementing design, decision making and modelling techniques in working production systems [197]. The programme area relates to the total range of activities in computer integrated manufacturing, including Computer Aided Design (CAD), Computer Aided Engineering (CAE), Computer Aided Manufacture (CAM), flexible machining and assembly systems, robotics, testing and quality control.

The objectives are to create an environment in which multi-vendor systems can be implemented, based on Open Systems concepts. The programme supports particularly the development of structured methods and interactive support tools, and the work is divided into the following R & D topics.

CIM architecture and communications A new generation of distributed parallel computing control systems is necessary which will meet the following requirements:

- Distributed systems with multilevel control architecture
- Knowledge based system support to assist normal and abnormal system behaviour
- Integration of intelligent sensors into the system architecture
- Reliable communications in the time-critical and hazardous applications of direct shop floor control.

The CIM strategy emphasises the development of standards and technology for multi-vendor systems, using architectures, design rules and interfaces conforming to the Open Systems Interconnection model.

Product design and analysis systems Design needs to become a more integral part of the whole CIM process, and a product model should be the nucleus of future design and analysis systems. An aim of CAD must be to reduce the design and production cycle time, which requires close integration between the process model and the pre-production planning activities. AI techniques, fast parallel processing and emergent techniques to handle large data and knowledge bases will be required.

Integrated Planning and Control Systems The objective is to develop and demonstrate integrated management and control systems, for example, to integrate order intake, production planning and scheduling with real time control of plant, machines and robots. Of particular interest will be the development of a dynamic scheduling and process planning system that would generate manufacturing schedules based on higher and lower level constraints as they become apparent in real time. Higher level constraints include those of product design, order priorities, product mix, machine and material resources. Lower level constraints include actual or predicted machine failure, loss of product quality and operator error. The system will have predictive abilities to allow rescheduling before changes in requirements or machine failure occur.

Robotics and shop floor systems   The integration of robots and handling systems for materials, parts and tools is required to achieve fully automated systems. The aims also include the development of modular tools, integrated into higher level systems, for fault prediction, diagnosis of causes and corrective action planning. AI techniques will be used to deal with uncertainties and to build up both operational and predictive knowledge from historical logging and observed effects of operational states. The aim should be operation in real time to maximise system reliability and availability, and to incorporate plant maintenance requirements in planning and scheduling systems.

   The programme also refers to the production of generic CIM models and their evaluation, and to the development of techniques for the dynamic transformation of a function-oriented into a flow-oriented organisation.

References

1 GINI, G., GINI, M., and MORPURGO, R.: 'A knowledge-based consultation system for automatic maintenance and repair', *in* ELLIS, T.M.R., and SEMENKOV, O.I. (Eds): 'Advances in CAD/CAM' (North Holland, 1983) pp. 495–505

2 GERSHWIN, S.B., HILDEBRANT, R.R. SURI, R., and MITTER, S.K.: 'A control theorist's perspective on recent trends in manufacturing systems', Procs 23rd IEEE Conf. on Decision and Control, Las Vegas, Dec. 1984

3 CHANG, T.C., and WYSK, R.A.: 'An introduction to automated process planning systems' (Prentice Hall, New Jersey, 1985)

4 ROOBECK, A., and ABBING, M.R.: 'The international implications of computer integrated manufacturing', *Int. Jl. Compr. Integd. Manufactg.*, 1988, **1**, pp. 3–12

5 LERNER, E. J.: 'Computer-aided manufacturing', *IEEE Spectrum*, Nov. 1981, pp. 34–39

6 'Advanced Manufacturing Technology: The impact of new technology on engineering batch production.' Advanced Manufacturing Systems Group, National Economic Development Office (NEDO), London, May 1985

7 LAWRENCE, A.: 'Management philosophy—same goal: different paths' Computers in Manufacturing 5, *Financial Times*, London, 2 June 1987, p. 25

8 CROSSLEY, T.R.: 'Manufacturing systems', *in* NICHOLSON, H. (Ed.): 'Modelling of dynamical systems'; Vol 2 (IEE, London, 1981)

9 GARDNER, L.B. (Ed.): 'Automated manufacturing' (ASTM, Philadelphia, 1985)
  (*a*) WATER, C.: 'Manufacturing process control specification with functional programming', pp. 125–135
  (*b*) HOPP, T.H., and LAU, K.C.: 'A hierarchical model-based control system for inspection', pp. 169–187
  (*c*) NILSSON, N.T.: 'Functions and dimensions of distributed assembly automation: an analytical networking approach', pp. 207–228
  (*d*) MACKULAK, G.T.: 'An examination of the IDEFo approach used as a potential industry standard for production control system design', pp. 136–149.

10 REMBOLD, U., BLUME, C., and DILLMANN, R.: 'Computer-integrated manufacturing technology and systems' (Marcel Dekker, NY, 1985)

11 COALES, J.F.: 'Control and automation', *Proc. IEE*, 1966, **113**, pp. 161–168

12 WISTREICH, J.G.: 'Automation in the iron and steel industry', UK Automn Council (UKAC), 1967, **S2**, pp. 1–21

13 NICHOLSON, H.: 'Applications of modern control theory'. Inaugural lecture, Univ. of Sheffield, UK, April 1968

14 'Effective CADCAM '87, Towards Integration'. I.Mech.E. Conf., London, 1987
  (*a*) BLACK, I., and MURRAY, J.L.: 'Conceptualization and computer aided design—an applications perspective', pp. 49–56.
  (*b*) SHENTON, A.T.: 'Computer integration of geometric modelling and engineering design systems by rule based programming', pp. 57–67
  (*c*) CLARKE, M.J.: 'On the problems involved in turning a computer aided draughting package into a computer aided design system', pp. 13–19
  (*d*) LANSIAUX, P.: 'CADCAM in Rolls-Royce aero engines', pp. 103–112
  (*e*) JACKSON, R.H.: 'Achieving effective CADCAM, Ten years of experience at Baker Perkins'
  (*f*) SMITH, P.J.: 'CADCAM data exchange—what it is and how to make it happen', pp. 135–140

(*g*) SEBBORN, M.J.: 'The selection and implementation of CADCAM within automotive component supply companies', pp. 75–80
15 WEBER-BROWN, N.E.: 'The application of computers to steelworks production controls', *Trans. Soc. Instrumn. Techy.*, Dec. 1966, pp. 222–231
16 'Optimal systems planning', Proc. IFAC Sympm., Case Western Reserve Univ., Cleveland, Ohio, IEEE, June 1968
    (*a*) WOOD, A.J.: 'Planning of large systems', pp. 154–164
    (*b*) BERRA, P.B., and BARASH, M.M.: 'The automated planning and optimisation of manufacturing processes'
17 Proc. 6th Int. Conf. on Flexible Manufacturing Systems, Turin, Italy, 1987, IFS Publicns., UK
    (*a*) NILSSON, K.: 'Adaptive tool systems', pp. 235–245
    (*b*) PERERA, D.T.S.: 'Simulation of tool flow within a flexible manufacturing system', pp. 211–222
    (*c*) BURSZTYN, D.: 'Integration of the plant's MRP II systems with the FMC's controls', pp. 53–67
    (*d*) HUBER, A.: 'Knowledge-based production control for a flexible flow line in a car radio manufacturing plant', pp. 3–19
    (*e*) ERCOLE, M.: 'FMS: zero QC v/s advanced integrated QC; is "zero defects" really possible?', pp. 301–308.
    (*f*) GUIDUCCI, A., SARACCO, O., and RUTELLI, G.: 'On line automatic detection and evaluation of the decayed area of a cutting tool', pp. 329–338
18 Proc. Int. Conf. on The Development of Flexible Automation Systems. IEE Conf Publicn 237, 1984, London
    (*a*) STEPHENS, A.P.: 'Tool management within a flexible manufacturing system', pp. 73–79
    (*b*) RANKY, P.G.: 'Pallet alignment error correction in FMS by means of software', pp. 93–98.
    (*c*) PERCIVAL, N.: 'The role of standards in flexible automation systems', pp. 52–54
    (*d*) KAY J.M.: 'The ue of modelling and simulation techniques in the design of manufacturing systems', pp. 55–61.
    (*e*) BURROW, L.D.: 'The 'Design to Product' Alvey demonstrator', pp. 114–123
19 CONSTANTINIDES, N, and BENNETT, S.: 'An investigation of methods for on-line estimation of tool wear', *Int. J. Mach. Tools Manufact.*, 1987, **27**, pp. 225–237
20 JARVIS, J.W.: 'Computer integrated manufacturing technology and status'—an overview for management'. Advanced Manufactg Informn., IEE, London, 1986
21 CHOW, W.-MIN, MACNAIR, E.A., and SAUER, C.H.: 'Analysis of manufacturing systems by the research queueing package', *IBM J. Res. Develop.*, 1985, **29**, pp. 330–341
22 HAINES, C.L.: 'An algorithm for carrier routing in a flexible material-handling system', *IBM J. Res. Develop.*, 1985, **29**, pp. 356–362
23 BROWNE, J., DUBOIS, D., RATHMILL, K., SETHI, S.P., and STECKE, K.E.: 'Classification of flexible manufacturing systems', *The FMS Magazine*, April 1984, pp. 114–117
24 Proc. 2nd UMIST/ACME Workshop Advanced Research in Computer Aided Manufacturing, Jan. 1987, UMIST, Manchester
    (*a*) WILLIAMS, D.J., and ROGERS, P.: 'Opportunities for knowledge based systems in the control of manufacturing automation'
    (*b*) ROWE, G., and HARTLEY, P.: 'Advanced research in computer-aided manufacturing. An expert system for designing forging dies'
    (*c*) DARBYSHIRE, I.L., WRIGHT, A. J., and DAVIES, B.J.: 'Development of EXCAP: An intelligent knowledge-based system for generative process-planning'
    (*d*) LIU, Y.S., and ALLEN, R.: 'A flexible, interactive front-end for a computer-aided process planning system'
    (*e*) HUSBANDS, P., MILL, F.G., and WARRINGTON, S.W.: 'Process planning: knowledge representation'
25 KUSIAK, A. (Ed.): 'Modelling and design of flexible manufacturing systems' (Elsevier, Amsterdam, 1986)
    (*a*) VAN LOOVEREN, A.J., GELDERS, L.F., and VAN WASSENHOVE: 'A review of FMS planning models', pp. 3–31
    (*b*) TURBAN, E., and SEPEHRI, M.: 'Applications of decision support and expert systems in flexible manufacturing systems', pp. 369–386

(c) BEN-ARIEH, D.: 'Knowledge based control system for automated production and assembly', pp. 347–368

(d) BARD, J.F.: 'The evolution of robotics in manufacturing', pp. 33–63

(e) MARTINEZ, J., ALLA, H., and SILVA, M.: 'Petri nets for the specification of FMSs', pp. 389–406

26 Proc. 3rd Int. Conf. Simuln. in Manufacturing, Turin, Italy, Nov, 1987, IFS Publs., UK

    (a) KEHOE, D.F., LITTLE, D., and WYATT, T.M.: 'An approach to FMS design for improved integration', pp. 263–278

    (b) BASTOS, J.M., and SHIRES, N.: 'Factory simulation for evaluating planning and control', pp. 51–66

27 V. DUNGERN, D., and SCHMIDT, G.: 'A new scheme for on-line fault detection and control of assembly operation via sensory information'. Proc. 25th IEEE Conf. on Decision and Control, Athens, 1986

28 HITOMI, K.: 'Manufacturing systems engineering' (Taylor and Francis, London, 1979)

29 CIM Europe 1987 Conference, Knutsford, UK, May 1987

    (a) ARLABOSSE, F.: 'Knowledge based systems for manufacturing'

    (b) GREENWOOD, N.R.: 'CIM—what will be its impact during the next few years'

    (c) RUSSELL, P.J.: 'Open systems architecture for CIM'

    (d) PFEIFER, T., and KOMISCHKE, M.: 'Low level communications systems for sensors'

    (e) HARHEN, J., COHEN, P., GRAVES, R., and KETCHAM, M.: 'Using multiple perspectives in manufacturing macro-planning'

30 LUH, J.Y.S.: 'An anatomy of industrial robots and their controls', *IEEE Trans.* 1983, **AC–28**, pp. 133–153

31 KOIVO, A.J., and GUO, T.H.: 'Adaptive linear control for robotic manipulators', *IEEE Trans.*, 1983, **AC–28**, pp. 162–171

32 CORTI, P., GINI, G., and GINI, M.: 'Software features for intelligent industrial robots', *Kybernetes*, 1979, **8**, pp. 149–154

33 SIMON, H.A.: 'The architecture of complexity', *Proc. Amer Philos. Soc.*, 1962, **106**, pp. 467–482

34 KIMEMIA, J., and GERSHWIN, S.B.: 'An algorithm for the computer control of a flexible manufacturing system', *IIE Trans*, 1983, **15**, pp. 353–362

35 NICHOLSON, H.: 'Structure of interconnected systems' (Peter Peregrinus, IEE, London, 1978)

36 VILLA, A., and ROSSETTO, S.: 'On the joint problem of dynamic part routing and station service control in flexible manufacturing systems', *Material Flow*, 1985, **2**, pp. 97–110

37 McCLEAN, C., MITCHELL, M., and BARKMEYER, E.: 'A computer architecture for small-batch manufacturing', *IEEE Spectrum*, May 1983, pp. 59–64

38 MESAROVIC, M.D., LEFKOWITZ, I., and PEARSON, J.D.: 'Advances in multi-level control'. IFAC Sympm, Tokyo, 1965

39 LASDON, L.S., and SCHOEFFLER, J.D.: 'Decentralized plant control'. Syst. Res. Centre, Case Inst Techy., Cleveland, Ohio, Research Report, *circa* 1966

40 PEARSON, J.D.: 'Multi-level control systems', Syst. Res. Centre, Case Inst Techy., Cleveland, Ohio, Research Report, *circa* 1967

41 COALES, J.F.: 'The modelling and control of large dynamic systems'. Research Report, 1972, Control Group, Eng. Dept., Univ. of Cambridge

42 ATHANS, M.: 'Advances and open problems on the control of large scale systems'. IFAC, Helsinki, Finland, 1978

43 ROSS, D.T.: 'Structured analysis (SA): A language for communicating ideas', *IEEE Trans.* 1977, **SE–3**, pp. 16–34

44 MATHER H: 'Logistics in manufacturing—the new competitive weapon', *I.Mech.E. Manufactg Bull*, Sept. 1987, (11)

45 BURBIDGE, J.L.: 'The simplification of material flow systems', *Int J. Prodn. Res.*, 1982, **20**, pp. 339–347

46 OTOLORIN, O.: 'Structural modelling and control of job-flow in manufacturing systems'. Ph.D. Thesis, Dept. of Control Eng., Univ. of Sheffield, UK, 1982

47 PARNABY, J.: 'Competitiveness via total quality of performance', *Progress in Rubber and Plastics Techy.*, 1987, **3**, pp. 42–50

48 BERTOK, P., CSURGAI, G., and HAIDEGGER, G.: 'Flexible manufacturing systems with general, easily reconfigurable cells'. Control 88, IEE Conf Publ. 285, pp. 207–211

152 References

49 KUSIAK, A: 'Artifical intelligence and CIM systems', in KUSIAK, A. (Ed.): 'Artificial intelligence, implications for CIM (IFS (Publns.), Springer Verlag, 1988) pp. 3–23
50 NICHOLSON, H., and LYNN, J.W.: 'The economic loading of transmission systems', *Proc. IEE*, Pt. C, 1958, **105**, pp. 407–419
51 BUZACOTT, J.A., and SHANTHIKUMAR, J.G.: 'Models for understanding flexible manufacturing systems', *Trans. AIIE*, 1980, **12**, pp. 339–350
52 RANKY, P.G.: 'The design and operation of FMS' (IFS (Publns.) Ltd., North Holland, 1983)
53 ELLIS, T.M.R., and SEMENKOV, O.I., (Eds): 'Advances in CAD/CAM'. Proc. 5th Int. IFIP/IFAC Conf. 1982, Prolamat 82, (North Holland, 1983)
   (*a*) TSVETKOV, V.D.: 'System-structural models and formalized language used to describe discrete manufacturing processes in the computer-aided design system', pp. 119–133
   (*b*) LEMOINE, Y., and MUTEL, B.: 'Automatic recognition of production cells and part families', pp. 239–248
54 HANNAM, R.G.: 'Alternatives in the design of flexible manufacturing systems for prismatic parts', *Proc. I.Mech.E.*, 1985, **199**, (B2), pp. 111–119
55 STECKE, K.E.: 'Useful models to address FMS operating principles'. Proc. IFIP Conf Advances in Prodn. Manag. Syst., Budapest, 1985
56 HARTLEY, J.: 'FMS at work', (IFS (Publns.) Ltd. North Holland, UK)
57 THOMPSON, J.: 'CIM: bringing the islands together', Chartered Mech. Eng., March 1987, pp. 49–52
58 DALBY, R: 'Implementing flexible automation', *Electronics & Power*, Aug. 1987, pp. 514–516
59 Ingersoll Engineers, 'Integrated manufacture' (IFS (Publns), 1985)
60 'Current and future trends of manufacturing management and technology in the UK.' Inst. of Prodn. Engrs., 1985
61 KOCHAN, A: 'DEC announces commitment to CIM', *FMS Magazine*, 1985, **3**, pp. 97–99
62 FEBVRE, K.L.: 'Communications hierarchy for the factory', *Electronics & Power*, March 1987, pp. 192–194
63 PLATT, E.: 'MAP and GEC', *GEC Review*, 1987, **3**, pp. 107–114
64 CORNWELL, P.: 'Implementing MAP/EPA in the manufacturing cell'. ESPRIT-CIM-Europe Spec. Int. Grp., No 1, Cells, Archits, and Commns., CIM-Europe Mtg., Dec. 1986 (Renishaw Controls Ltd)
65 BISHOP, R.E.D.: 'On the teaching of design in universities'. I.Mech.E. Paper 27/63, Educn. and Training Group, London, 1963
66 IEE/Design Council Wkg. Party Report: 'The teaching of design'. IEE, London, 1983
67 FRENCH, M.J.: 'Conceptual design for engineers'. The Design Council, London, (Springer Verlag, Berlin, 1985)
68 PAHL, G., And BEITZ, W.: 'Engineering Design' (Edited by K WALLACE). The Design Council, London, (Springer Verlag, Berlin, 1984)
69 Report of the Engineering Design Working Party, SERC, UK, 1983
70 BJORKE, O., and FRANKSEN, O.I. (Eds.): 'System structures in engineering: economic design and production'. TAPIR, 1978
   (*a*) YOSHIKAWA, H.: 'Multipurpose modelling of mechanical systems—morphological model as a mesomodel', pp. 594–629
   (*b*) FALSTER, P., and ROLSTADAS, A.: 'Interactive planning of lot sizes for product mix in manufacturing cells', pp. 534–593
71 LEE, S.S.G., and ATKINSON, J.: 'Integration of metals selection with design analysis', *Chartered Mech. Eng*. Oct, 1987, pp. 35–38.
72 BYRNE, D.M., and TAGUCHI, S.: 'The Taguchi approach to parameter design'. Amer. Assocn. for Quality Control, Quality Congress Trans., Anaheim, 1986, pp. 168–177
73 BURGAM, P.M.: 'Design of experiments—the Taguchi way', *Manufactg. Eng.*, May 1985, pp. 44–47
74 Proc. UMIST/ACME Workshop on Process Planning Automation and Analysis, UMIST, UK, Dec. 1985
   (*a*) HANNAM, R.G.: 'Automated process planning through CADCAM', p. 91–95
   (*b*) ISMAIL, H.S., and HON, K.K.B.: 'An expert system for the selection of machining processes', pp. 83–90
   (*c*) STEWART, J.J.T.: 'Generative process planning and its wider implications', pp. 27–37
   (*d*) FIRTH, P.: 'AI techniques for automated process planning', pp. 120–122

(e) BENNATON, J., CASE, K., ACAR, S., and HART, N.: 'A CAD system to aid planning for manufacture', pp. 123–131

(f) WOODHEAD, R., and SAIA, A.: 'Towards an intelligent planning system', pp. 7–12

(g) ROGERS, P., CHU, E., and WILLIAMS, D.J.: 'Opportunities for the use of planners in the control of manufacturing cells', p. 49–66

(h) SWIFT, K.G., SYAN, C.S., and MATTHEWS, A.: 'A logic-based approach to problem solving in manufacturing engineering', pp. 72–82

75 SWIFT, K.G.: 'Knowledge-based design for manufacture', (Kogan Page, London, 1987)

76 ACME Workshop on Expert Systems in Manufacturing Engineering, SERC, Nottingham, UK, 1984

(a) SWIFT, K.G., MATTHEWS, A., and RUNCIMAN, C.: 'Knowledge based systems in engineering design'

(b) PARKINSON, A.: 'Expert systems and geometric modelling'

(c) DUFFY, A., and MacCALLUM, K.: 'The Designer system'

(d) DARBYSHIRE, I., and DAVIES, B.J.: 'EXCAP: An expert generative process planning system'

77 PITTS, G., and HODGSON, B.A.: 'Design for economic manufacture—designing to minimise transfers between machine tools'. Proc. Research Conf., ACME, Cambridge Univ., 1987

78 BERTRAND, J.W.M., and WORTMANN, J.C.: 'Production control and information systems for component-manufacturing shops', (Elsevier, Amsterdam, 1981)

79 RUSHTON, D.F.H., HODGKINSON, G.D., BROUGHTON, T., and WINSTONE, J.: 'A guide to manufacturing strategy' Prodn. Engrs., London, 1986

80 MERCHANT, M.E.: 'Technological forecasting—an essential component of today's technology', *Int. J. Systs Sci.*, 1970, **1**, pp. 49–62

81 CLOWES, G.A., and MARSHALL, S.A.: 'Corporate financial models: a survey', *Proc. IEE*, 1972, **119**, pp. 369–376

82 NICHOLSON, H.: 'Sequential least-squares prediction based on spectral analysis', *Int. J. Compr. Maths.* Sectn B, Comptl Methods, 1972, **3**, pp. 257–270

83 FERNANDO, K.V.M., and NICHOLSON, H.: 'Two-dimensional curve-fitting and prediction using spectral analysis', *Proc. IEE*, 1982, **129**, pp. 145–150.

84 STECKE, K.E.: 'Design, planning, scheduling, and control problems of flexible manufacturing systems', *Ann. Operns. Res.*, 1985, **3**, pp. 3–12

85 BESSANT, J.: 'Technology and market trends in the production and application of information technology. A review of developments during the years 1982–83'. UNIDO Report UNIDO/IS, 438, 8, Feb. 1984

86 GERSHWIN, S.B.: 'Material and information flow in an advanced automated manufacturing system'. Report LIDS-P-1199. Laby. Informn. and Decisn. Systs., Mass. Inst. Techy., May 1982

87 O'GRADY, P.J.: 'Putting the Just-in-Time philosophy in practice: a strategy for production managers' (Kogan Press, 1988)

88 BIELECKI, T., and KUMAR, P.R.: 'Optimality of zero-inventory policies for unreliable manufacturing systems'. Report, Dept. of Elect. and Compr. Eng. and Coordd. Science Laby., Univ of Illinois, USA

89 BASTOS, J.M.: 'Batching and routing: two functions in the operational planning of flexible manufacturing systems', *Europn. J. Operl. Res.*, 1988, **33**, pp. 230–244

90 VILLE, A.: 'Hierarchical architectures for production planning and control'. Advanced School on Operns. Res. Models in Flex. Manufactg. Systs., CISM, Udine, Italy, Oct. 1987

91 CHANG, T.C., and WYSK, R.A.: 'An integrated CAD/automated process planning system', *Trans. AIIE*, 1981, **13**, pp. 223–233

92 EVERSHEIM, W., and ESCH, H.: 'Automatic generation of process plans for prismatic parts', *Ann. CIRP*, 1983, **32**, pp. 361–364

93 CARRIE, A.S., and PERERA, D.T.S.: 'Work allocation in flexible manufacturing systems', Conf. Compr Aided Prodn. Eng., March 1986, Edin., UK, pp. 1–5

94 CARRIE, A.S., and PERERA, D.T.S.: 'Work scheduling in FMS' Conf Opertl. Res. Soc., Durham, UK, 1985, pp. 1–8.

95 MENON, U., and O'GRADY, P.J.: 'A flexible multiobjective production planning framework for automated manufacturing systems'. Eng. Costs and Prodn Econs., 1984

96 LUCERTINI, M., and NICOLO, F.: 'Workstation set-up in FMS'. Workshop, Operns Research Models in Flex. Manufactg. Systs., Udine, Italy, Oct. 1987.

97 NEWMAN, P.A.: 'Scheduling in CIM systems', *in* KUSIAK, A. (Ed.); 'Artificial intelligence implications for CIM'. (IFS Publicns, Springer Verlag, 1988), pp. 361–402

98 KANET, J.J., and ADELSBERGER, H.H.: 'Expert systems in production scheduling', *Europn. J. Operl. Res.*, 1987, **29**, pp. 51–59

99 ASHOUR S: 'Sequencing theory'. Lect. notes in Econs, and Mathl. Syst. (Springer Verlag, Berlin, 1972)

100 DOUMEINGTS, G., DUMORA, E., CHABANAS, M., and HUET, J.F.: 'Use of GRAI method for the design of an advanced manufacturing system'. Proc. 6th Int. Conf. Flexible Manufactg. Systs., 1987, Nov. IFS Confs., pp. 341–358

101 ROLSTADAS, A.: 'Scheduling batch production by means of an on-line mini computer'. 2nd Int. Conf. Compr. Aided Manufacture, East Kilbride, Scotland, June 1978

102 LENZ, J.E.: 'General theories of flexible integration'. Proc. 5th Int. Conf. Flexible Manufactg. Syst., 1986, IFS (Confs.), pp. 255–264

103 CHAN, T.S., and PAK, H.A.: 'Heuristical job allocation in a flexible manufacturing system', *Int J. Advanced Manufactg Techy.*, 1986, **1**, pp. 69–90

104 FALSTER, P., and ROLSTADAS, A.: 'The integration of production management and flexible manufacturing', *in* ELLIS, T.M.R., and SEMENKOV, O.I. (Eds.): 'Advances in CAD/CAM 83' (North Holland, 1983), pp. 303–313

105 MEYER, W., ISENBERG, R., and HÜBNER, M: 'Knowledge-based factory supervision—The CIM shell', *Int. J. Compr. Intgd. Manufactg.*, 1988, **1**, pp. 31–43

106 BRANDIMARTE, P., CONTERNO, R., and LAFACE, P: 'FMS production scheduling by simulated annealing'. Proc. 3rd Int. Conf. Simuln. in Manufactg., 1987, IFS (Confs), pp. 235–245

107 ROGERS, P., and WILLIAMS, D.J.: 'Knowledge-based control in manufacturing automation', *Int. J. Compr. Intgd. Manufactg.*, 1988, **1**, pp. 21–30

108 BAKER, K.R.: 'Introduction to sequencing and scheduling' (Wiley, NY, 1974)

109 PANWALKAR, S.S., and ISKANDER, W: 'A survey of scheduling rules', *Operns. Res.*, 1977, **25**, pp. 45–61

110 WITTROCK, R.J.: 'Scheduling algorithms for flexible flow lines', *IBM J. Res. Develop.*, 1985, **29**, pp. 401–412

111 FOX, M.S., and SMITH, S.: 'ISIS—a knowledge-based system for factory scheduling', *Expert Systems*, 1984, **1**, pp. 25–47

112 FOX, M.S., and SMITH, S.: 'Constraint-based scheduling in an intelligent logistics support system: An artificial intelligence approach. Intelligent Systems Laby. Report, 1985, Robotics Institute, Carnegie–Mellon Univ., Pittsburgh, USA

113 SMITH, S.: 'The use of multiple problem decompositions in time constrained planning tasks'. Proc. Int. Joint Conf. on Artificial Intell., 1985, pp. 1013–1015

114 SMITH, S.F., and FOX, M.S.: 'Constructing and maintaining detailed production plans: Investigations into the development of knowledge-based factory scheduling systems'. Report 1985, Carnegie–Mellon Univ., Pittsburgh, USA

115 WALKER, T.C., and MILLER, R.K.: 'Expert system 1986: An assessment of technology and applications'. SEAI Technl Publns, 1986, Madison, USA

116 CHARALAMBOUS, O., and HINDI, K.S.: 'Artificial intelligence-based scheduling systems—A review and annotated bibliography', Report No 88/12, Decision Techns. Group of Comput. Dept., UMIST, UK

117 WANG, L.: 'Robustness of adaptive control systems'. Ph.D. Thesis, Dept. of Control Eng., Univ of Sheffield, UK, 1988

118 DAVISON, E.J.: 'Multivariable tuning regulators: the feedforward and robust control of a general servomechanism problem', *IEEE Trans.*, 1976, **AC-21**, pp. 35–47

119 OWENS, D.H., and CHOTAI, A.: 'Robust controller design for linear dynamic systems using approximate models' *Proc. IEE*, Pt. D, 1983, **130**, pp. 45–56

120 HARRIS, C.J., and BILLINGS, S.A. (Eds): 'Self-tuning and adaptive control: theory and applications' (Peter Peregrinus, IEE, London, 1981)

121 McGOLDRICK, P.F., HIJAZI, M.A.M., and GEDAY, A.: 'Optimization strategies for adaptively controlled milling', *in* ELLIS, T.M.R., and SEMENKOV, O.I. (Eds.): 'Advances CAD/CAM 83', (North Holland, 1983), pp. 211–219

122 SELIGER, G., VIEHWEGER, B., and WIENEKE, B: 'Decision support in design and optimization of flexible automated manufacturing and assembly'. Int. Conf. Intell. Manufactg. Syst., Budapest, 1986

123 KORIBA, M.: 'System modelling for the competitive edge', *Chartered Mech. Eng.*, April 1988, pp. 38–39.

124 TOWILL, D.R., EDGHILL, J., and JONES, M.: 'Improved manufacturing system design via industrial dynamics models'. Proc. Research Conf., ACME, Cambridge Univ., 1987

125 CHRISTENSEN, J.L., and BROGAN, W.L.: 'Modelling and optimal control of a production process', *Int. J. Systs. Sci.*, 1971, **1**, pp. 247–255

126 WIESLAW, N.: 'Modelling of discrete production systems'. Proc. 2nd Int. Conf. Syst. Eng., Coventry, Lanchester Poly., UK, Sept. 1982, pp. 142–147

127 POWELL, N.K.: 'Modelling production systems—a general systems approach', *in* KUSIAK, A. (Ed.): 'Modern Prodn. Manag. Systs' (Elsevier, North Holland, 1987)

128 CROOKALL, J.R.: 'Computer integration of advanced manufacture', *Proc. I.Mech.E.*, 1986, **200**, (B4), pp. 257–264

129 SPUR, H.C., HIRN, W., and SELIGER, G.: 'The role of simulation in design of manufacturing systems'. Proc. IFIP/IFAC Prolamat 1982, Leningrad

130 BONNEY, M., and BYRNE, M.: 'Batch sizing and flexible manufacture'. Proc. Research Conf, ACME, SERC, 1987

131 WILLKE, T.L., and MILLER, R.A.: 'Production planning and control—II. The planning horizon problem', *Int. J. Systs. Sci.*, 1978, **9**, pp. 1259–1270

132 GORDON, W., and NEWELL, G.: 'Closed queueing systems with exponential servers', *Oper. Res.*, 1967, **15**

133 BUZEN, J.: 'Computational algorithms for closed queueing networks with exponential servers', *Commun. ACM*, 1973, **16**

134 KIMEMIA, J., and GERSHWIN, S.B.: 'Flow optimisation in flexible manufacturing systems', *Int. J. Prod. Res.*, 1985, **23**, pp. 81–96

135 SOLBERG, J.J.: 'Capacity planning with a stochastic workflow model', *J. AIIE*, 1981, **13**

136 STECKE, K.E., and SOLBERG, J.J.: 'Optimality of unbalanced workloads and machine group size for flexible manufacturing systems', Working paper 290, Div. Res., Graduate School of Bus. Admin., Univ of Michigan, 1982

137 SURI, R.: 'New techniques for modelling and control of flexible automated manufacturing systems'. Proc. 1981 IFAC, Kyoto, Japan

138 SURI, R: 'Robustness of queueing network formulas', *J. ACM*, 1983, **30**, pp. 564–594

139 YAO, D.D.: 'An FMS network model with state-dependent routing'. Proc 1st ORSA/TIMS Conf. on FMS, Ann Arbor, 1984

140 SHALEV-OREN, S., SEIDMAN, A., and SCHWEITZER, P.: 'Analysis of flexible manufacturing systems with priority scheduling: PMVA', Proc. 1st ORSA/TIMS Conf. on FMS, Ann Arbor, 1984

141 SOLBERG, J.J., and NOF, S.Y.: 'Analysis of flow control in alternative manufacturing configurations'. *J. Dyn. Systs. Meas. and Control, Trans. ASME*, 1980, **102**, pp. 141–147

142 COHEN, G., DUBOIS, D., QUADRAT, J.P., and VIOT, M.: 'A linear-system theoretic view of discrete-event processes', Proc. 22nd IEEE Conf. Decisn. and Control, Dec. 1983

143 COHEN, G., DUBOIS, D., QUADRAT, J.P., and VIOT, M.: 'A linear-system-theoretic view of discrete-event processes and its use for performance evaluation in manufacturing', *IEEE Trans.*, 1985, AC-30, 3, pp. 210–220.

144 HO, Y-C.: 'Performance evaluation and perturbation analysis of discrete event dynamic systems', *IEEE Trans.* 1987, **AC-32**, pp. 563–572

145 SURI, R., and DILLE, J.W.: 'On-line optimization of flexible manufacturing systems using perturbation analysis'. Proc. 1st ORSA/TIMS Conf. on Flex. Mfg. Systs., Aug. 1984

146 CASSANDRAS, C.G., and STRICKLAND, S.G.: 'Perturbation analytic methodologies for design and optimisation of communication networks'. Tech. Report, Dept. of Elect. and Compr. Eng., Univ. of Massachusetts, 1987

147 HO, Y.C., SURI, R., CAO, X.R., DIEHL, G.W., DILLE, J.W., and ZAZANIS, M.A.: 'Optimization of large multiclass (non-product-form) queueing networks using perturbation analysis'. Large scale systems, 1984

148 HO, Y.C., and CASSANDRAS, C.: 'Computing costate variables for discrete event systems'. Proc. 19th IEEE Conf. Decisn. and Control, Dec. 1980

149 HO, Y.C.: 'On the perturbation analysis of discrete-event dynamic systems', *J. Optmn. Theory and Applns.*, 1985, **46**, pp. 535–545

150 HO, Y.C., and CASSANDRAS, C.: 'A new approach to the analysis of discrete event dynamic systems', *Automatica*, 1983, **19**, pp, 149–167

151 CASSANDRAS, C.G.: 'On-line decision aids in computer integrated manufacturing'. IFAC Conf, Compr. Aided Design in Control and Engg. Systems, Lyngby, Denmark, 1985, pp. 381–385

152 STECKE, K.E.: 'Formulation and solution of nonlinear integer production planning problems for flexible manufacturing systems', *Management Science*, 1983, **29**, pp. 273–288

153 KIMEMIA, J.G.: 'Hierarchical control of production in flexible manufacturing systems'. Ph.D. Dissertn., MIT Lab. Informn. and Decision Syst., Report LIDS-TH-1215, 1982

154 STECKE, K.E., and TALBOT, F.B.: 'Heuristic loading algorithms for flexible manufacturing systems'. Proc. 7th Intl. Conf. on Prodn. Research, Windsor, Canada, 1983

155 MESAROVIC, M.D., MACKO, D., and TAKAHARA, Y: 'Theory of hierarchical, multilevel systems' (Academic Press, NY, 1970)

156 SINGH, M.G., and TITLI, A.: 'Systems decomposition, optimisation and control' (Pergamon Press, Oxford, 1978)

157 CONTERNO, R., FIORIO, G., MENGA, G., and VILLA, A.: 'A large scale system approach to the production planning and control problem'. Proc 24th Conf. Decisn. and Control, Ft. Lauderdale, Fl., 1985, pp. 2016–2021

158 PERERA, D.T.S., and CARRIE, A.S.: 'The part selection in flexible manufacturing systems with high tool variety'. 3rd Int. Conf. Advances in Prodn. Manag. Systs. (APMS 87), Manitoba, Canada, 1987

159 GERSHWIN, S.B., AKELLA, R., and CHOONG, Y.F.: 'Short-term production scheduling of an automated manufacturing facility', *IBM J. Res. Develop.*, 1985, **29**, pp. 392–400

160 BIRLEA, S.: 'Entropy in industrial cybernetic systems', *Kybernetes*, 1973, **2**, pp. 85–93

161 PETERSON, J.L.: 'Petri net theory and the modelling of systems' (Prentice Hall, NJ, 1981)

162 PETERSON, J.L.: 'Petri nets', *Computing Surveys*, 1977, **9**, pp. 223–252

163 ARCHETTI, F., and SCIOMACHEN, A.: 'Representation, analysis and simulation of manufacturing systems by Petri net based models'. Proc. IIASA Conf. Discrete Event Systems: Models and Applicns., Sopron, Hungary, Aug. 1987

164 AGERWALA, T.: 'Putting Petri nets to work', *Computer*, Dec. 1979, pp. 85–94.

165 DUBOIS, D., and STECKE, K.E.: 'Using Petri nets to represent production processes'. Proc. 22nd IEEE Conf. Decisn. and Control, Dec. 1983, pp. 1062–1067

166 HEAD, M.A., BARSON, R.J., and BONNEY, M.C.: 'A Petri net representation of computer aided production management'. Research Report, Dept. of Prodn. Eng. and Prodn. Management, Univ. Park, Univ. of Nottingham, UK.

167 FARAH, B.N.: 'An expert support system for analysis and design of information systems', *in* KUSIAK, A. (Ed.): 'Artificial intelligence implications for CIM', (IFS (Publns.), Springer Verlag, 1988), pp. 435–457

168 BOBBIO, A., and SAVANT AIRA, G.: 'Modelling automated production systems by deterministic Petri nets'. Proc. 3rd Int. Conf. Simuln. in Manufactg., Nov 1987, IFS (Conf.), pp. 127–136

169 TANKEH, N.A.: 'Knowledge based techniques in manufacturing systems'. Project Report, Dept. of Control Eng., Univ of Sheffield, UK, 1989

170 BARASH, M.M., BARTLETT, E., FINFTER, I.I., and LEWIS, W.C.: 'Process planning automation—a recursive approach. The optimal planning of computerised manufacturing systems'. Report 17, School of Industl. Eng., Purdue Univ, USA, 1980

171 LEWIS, W.C., BARTLETT, E., FINFTER, I.I., and BARASH, M.M.: 'Tool-oriented process-planning'. Proc 23rd MTDR Conf., UMIST, Sept. 1982

172 BASDEN, A.: 'On the application of expert systems', *Int. J. Man–Machine Studies*, 1983, **19**, pp. 461–477

173 HALEVI, G., and WEILL, R.: 'Development of flexible optimum process planning procedures', *Ann. CIRP*, 1980, **29**, p. 313–317

174 MATSUSHIMA, K., OKADA, N., and SATA, T.: 'The integration of CAD and CAM by application of artificial-intelligence techniques', *Ann. CIRP*, 1982, **31**, pp. 329–332.

175 FOX, M. S.: 'Job shop scheduling: an investigation in constraint-directed reasoning', *Proc. NCAI*, 1982, pp. 155–158.

176 BUNCE, P.G.: 'The need for factory modelling'. CIM Europe 1987 Conf., Esprit, Knutsford, UK, May 1987

177 KUSIAK, A.: 'An expert system for group technology', *Industl. Engg.,* Oct. 1987, pp. 89–93

178 LEONARD, R.: 'Elements of cost-effective CIM', *Int. J. Compr. Intgd. Manufactg.,* 1988, **1**, pp. 13–20.

179 PLOSSL, G.W., and WIGHT, O.W.: 'Production and inventory control principles and techniques' (Prentice Hall, 1967)

180 ORLICKY, J.A.: 'Material requirements planning—the new way of life in production and inventory management' (McGraw Hill, 1975)

181 NEW, C.: 'Requirements planning', (Gower Press, 1973)

182 MEYER W: 'Knowledge-based realtime supervision in CIM—the workcell controller' *in* Directorate General XIII (Eds.) 'Esprit 86, Results and achievements', (Elsevier, North Holland, 1987)

183 'JITs first users present the facts', *Engineering News,* April 1987, (26)

184 OAKES, F.: 'The Japan syndrome. Part 2: The Japanese manufacturer and his society', *Electronics & Power,* Oct. 1987, pp. 619–623

185 VOSS, C.A.: 'Two sides of the JIT coin—Japan and UK', *Chartered Mech. Eng.,* April 1988, p. 29–31 Serial 621.

186 SHINGO, S.: 'Zero quality control: source inspection and the poka-yoke system' (Productivity Press, Stamford, USA, 1986)

187 BUTCHER, J.: 'Challenges to engineering education', *Proc. IEE,* 1984, **131**, Pt. A, pp. 662–664

188 BEER, S.: 'Platform for change', (Wiley, London, 1975)

189 HORTON, P., and NICHOLSON, H.: 'Generation of oscillatory behaviour in the Laisk model of photosynthetic carbon assimilation', *Photosynthesis Research,* 1987, **12**, pp. 129–143

190 GERSHWIN, S.B.: 'Editorial—Opportunities for control in manufacturing', *IEEE Trans.* 1985, **AC–30**, p. 833.

191 MURATA, T.: 'State equation, controllability, and maximal matchings of Petri nets', *IEEE Trans.,* 1977, **AC–22**, pp. 412–416

192 MURATA, T.: 'Petri nets' *in* SINGH, M.G. (Ed.): 'Systems and control encyclopedia: Theory, technology, applicns, Vol. 6 (Pergamon Press, Oxford, 1987), pp. 3665–3670

193 MURATA, T.: 'Circuit theoretic analysis and synthesis of marked graphs'; *IEEE Trans.,* 1977, **CAS–24**, pp. 400–405.

194 BROFFERIO, S.: 'A Petri net control unit for high speed modular signal processors'. Research Report, Dipartimento di Elettronica, Politecnico di Milano, Italy

195 CORSI, F., and CASTAGNOLO, B.: 'Probabilistic delay evaluation in combinational digital circuits by Petri nets', *Microelectron Reliab.,* 1983, **23**, pp. 541–553

196 SUZUKI, I., and MURATA, T.: 'A method for stepwise refinement and abstraction of Petri nets', *J. Computer and Syst. Science,* 1983, **27**, pp. 51–76

197 ESPRIT, European Strategic Programme for Research and Development in Information Technology, 1987 Annual Report, Commission of the European Communities

198 RANKY, P.G.: 'A real-time, rule-based FMS operation control strategy in CIM environment—Part I', *Int. J. Compr. Intgd. Manufactg.,* 1988, **1**, pp. 55–72

199 KUSIAK, A.: 'Designing expert systems for scheduling of automated manufacturing', *Industl. Engg.,* July 1987, pp. 63–67

200 ISENBERG, R.: 'Integrating the manufacturing process using expert system technology'. Artificial intelligence methods and tools in computer integrated manufacture, Workshop Proc., Esprit CIM Europe, SIG 'Advanced Informn Processg in CIM', Athens, Jan. 1987, pp. 163–186

201 ROGERS, P., and WILLIAMS, D.J.: 'Knowledge-based control in manufacturing automation', *Int. J. Compr. Intgd. Manufactg.,* 1988, **1**, pp. 21–30

Index